PLANTS

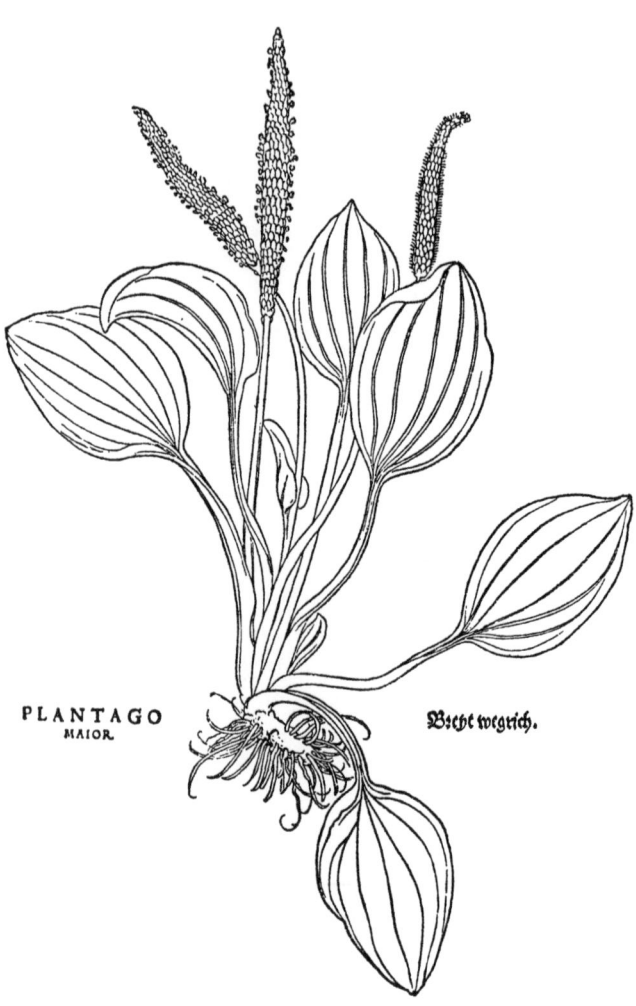

PLANTAIN. A reduced copy of a woodcut from the famous Herbal, *De Historia Stirpium*, written and illustrated by Leonhard Fuchs, Professor of Medicine at the University of Tübingen; published in 1542.

(From Mrs Arber's *Herbals*, Camb. Univ. Press.)

PLANTS
WHAT THEY ARE AND WHAT THEY DO

BY

A. C. SEWARD

F.R.S., Sc.D., D.Sc., LL.D.
*Master of Downing College &
Professor of Botany
Cambridge*

CAMBRIDGE
AT THE UNIVERSITY PRESS
1932

CAMBRIDGE UNIVERSITY PRESS
Cambridge, New York, Melbourne, Madrid, Cape Town,
Singapore, São Paulo, Delhi, Tokyo, Mexico City

Cambridge University Press
The Edinburgh Building, Cambridge CB2 8RU, UK

Published in the United States of America by
Cambridge University Press, New York

www.cambridge.org
Information on this title: www.cambridge.org/9781107600072

© Cambridge University Press 1932

This publication is in copyright. Subject to statutory exception
and to the provisions of relevant collective licensing agreements,
no reproduction of any part may take place without the written
permission of Cambridge University Press.

First published 1932
First paperback edition 2011

A catalogue record for this publication is available from the British Library

ISBN 978-1-107-60007-2 Paperback

Cambridge University Press has no responsibility for the persistence or
accuracy of URLs for external or third-party internet websites referred to in
this publication, and does not guarantee that any content on such websites is,
or will remain, accurate or appropriate.

What Nature has writ with her lusty wit
  Is worded so wisely and kindly
That whoever has dipped in her manuscript
  Must up and follow her blindly.

W. E. HENLEY

CONTENTS

		page
Preface		ix
Chap. I	Plants and Animals Compared	1
II	Response of Plants to Stimuli	9
III	The Superiority of Green Plants to Animals	22
IV	How a Plant obtains Energy	32
V	Cells and Tissues	39
VI	The Green Leaf	49
VII	The Green Leaf (*continued*): How a Leaf utilizes Radiant Energy	59
VIII	Roots and what they do	73
IX	Our Debt to Bacteria: the World's Supply of Nitrogen	83
X	Seeds and Seedlings	94
XI	Early Stages in the Evolution of Plants	110
XII	Later Stages in Evolution	123
Books suggested for further reading		131
Glossary		135
Index		139

PREFACE

My aim in writing this book is to present a few aspects of plant-life, in language as free as possible from technical terms, to readers who have little or no knowledge of botany or other branches of natural science. The book is not designed for students whose bent has caused them to specialize in science during the later stages of their school career, nor is it based on any examination schedule. The primary purpose is to introduce students whose school work, either by choice or the force of circumstances, is mainly classical or literary to some of the more striking attributes of the humbler organisms in order that they may be in a better position to appreciate what the world owes to plants. In more general terms, by describing what a plant is and what it does, I have attempted to awaken in laymen an interest in some of the fundamental principles of biology.

There is no lack of elementary text-books on botany, and though this book is not in any sense a text-book some reason for its publication should be given. I have been told on good authority that one of the chief obstacles in the path of masters and mistresses who would like to give every pupil an opportunity of obtaining some knowledge of Natural Science as an essential part of education, and not primarily as a preparation for examinations, is the dearth of suitable books. This is my excuse for venturing to offer a small contribution which may possibly be found helpful to teachers who realize the importance of responding to the urgent demand for creating an intelligent interest in natural phenomena and an appreciation of the meaning of scientific research.

In order so far as possible to meet the common and often well-deserved criticism that books described by authors as intelligible to the layman are rendered repellent by the frequent use of scientific jargon, I have explained in the text such technical terms as it seemed necessary to employ and for convenience of reference a glossary is appended (p. 135). I am grateful to Dr W. W. Vaughan, formerly Headmaster of Rugby, who very kindly read the typescript and encouraged me to offer it for publication, for suggesting both the addition of a glossary and a list of books (p. 131) suitable for readers who wish to make a closer acquaintance with subjects briefly dealt with in this sketch.

I cannot adequately express my indebtedness to my two colleagues Dr Maskell and Dr Godwin. Dr Maskell very kindly contributed the paragraphs, enclosed in quotation marks, on pages 55–56, and 77–80. I wish also to thank Mr W. Cuttle for checking the Greek and Latin derivations given in the Glossary.

A. C. SEWARD

Botany School, Cambridge
December 1931

COMMON ENGLISH EQUIVALENTS OF THE SCIENTIFIC UNITS OF MEASUREMENT

1 Metre	= 39·37 ins.; approximately 40 ins.
1 Centimetre	= $\frac{1}{100}$ of a metre; approximately $\frac{2}{5}$ of an inch
1 Millimetre	= $\frac{1}{1000}$ of a metre; approximately $\frac{1}{25}$ of an inch
1 μ (the Greek *m*)	= $\frac{1}{1000}$ of a millimetre; approximately $\frac{1}{25000}$ of an inch
1 Gram	= 15·43 grains; approximately $\frac{1}{28}$ of an ounce

CHAPTER I

Plants and Animals Compared

With few exceptions animals possess special organs of locomotion; their ability to travel on land, in air and water is one of the most obvious external signs of their aliveness. Plants such as most of us know are sedentary: their branches and leaves sway and flutter passively in the wind but the plant as a whole remains rooted in the ground. Plants appear to belong to a lower plane of existence and it is difficult to think of them as equally entitled with animals to be regarded as living organisms in the complete sense. In animals, movement is one of the most striking features; in plants, growth is the main thing. Rather more than two and a half centuries ago one of the first English botanists, Nehemiah Grew, said that plants and animals came at first out of the same hand and were therefore the contrivances of the same wisdom; that is as far as he went. At a still earlier date a French naturalist, Guy de Brosse, the founder of the Jardin des Plantes, Paris, who died in 1641, the date of Nehemiah Grew's birth, had convinced himself of the essential unity of vegetable and animal life. The average person thinks of plants as things of beauty, gems of Nature's handicraft created for our enjoyment and for our use: "Consider the lilies of the field, how they grow; they toil not, neither do they spin: and yet I say unto you, that even Solomon in all his glory was not arrayed like one of these". Plants seem to grow in some mysterious way and in due course reproduce their kind: most of us are content to admire their form and colour without enquiring into the nature of the processes by which the embryo within the seed is transformed into a tree or at the spring of the year there occurs "the miracle of earth reclad".

In the following pages we will consider some of the differences and resemblances between animals and plants and, in a later chapter, cite some of the abundant evidence in support of the French naturalist's view that there is no essential difference between the two kingdoms of the living world. We shall give reasons for believing that plants and animals are descended from a common ancestral stock, and that the great majority of plants occupy a position in one important respect superior to that held by the whole animal creation not excluding man. Nearly three centuries ago Sir Thomas Browne wrote in the *Religio Medici*: "'All Flesh is Grass' is not only metaphorically, but literally true; for all creatures we behold are but the herbs of the field, digested into flesh in them, or more remotely carnified in ourselves".

If we compare a human being with a tree it is difficult to find any real resemblance; both are alive; both pass through a period of youth before reaching maturity; both in course of time die and decay: but to compare more closely the life of an oak tree and the life of a man seems futile. A tree grows where the seed germinated; it remains fixed in the earth and year by year the stem grows higher, the number of branches and leaves increases, and the crown gradually expands. Each year in the growing season the tree adds to the girth of the stem, produces a fresh crop of buds, some destined to form leafy shoots, others to blossom and bear fruit and seed. A man grows for a relatively short time; when a certain stage is reached increase in height ceases and no new parts are developed. A tree, on the other hand, seems to retain its vigour for an almost unlimited period and enjoys perpetual youth. The stem and older branches show signs of old age and decrepitude while the yearly output of fresh shoots and flowers affords proof of unimpaired activity. The height to which a tree can grow is controlled by mechanical laws; there is a limit of stability. In comparison with men and the larger animals the size of a tree and its length of years are immense. It is well known that some of the largest of the big trees of California indicate an age of over 3000 years. Each year a fresh cylinder of wood is added to the stem, and by counting the rings, that are usually well defined on a smooth cross-section, it is easy to estimate the age of the tree with approximate accuracy. Though we cannot learn very much by measuring the relative breadth of the concentric cylinders of wood, it is possible in some degree to visualize the varying response of the living plant to the fluctuating conditions of its environment over a period which began at least a thousand years before the Christian era. Here we have one of the most striking differences between a tree and the higher animals.

An animal begins life as a minute ball of living substance, a microscopic piece of protoplasm which is the fertilized egg. Could we give a true and complete definition of protoplasm, what it is and how, exactly, it differs from other substances, we should know what life is. The egg, stimulated by the addition of the male germ, increases in volume, gradually producing more and more protoplasm:

the embryo becomes progressively more elaborate. From an apparently uniform mass consisting of a group of minute pieces, each similar to the original egg, it rapidly changes its form and structure and the simple embryo develops into a complex adult no longer capable of further expansion. The body reaches a condition of relative permanence; wounds may be healed but no new limbs or other members are formed.

Plants also begin life as simple pieces of protoplasm—the living foundation substance common to both kingdoms—and in essentials the earliest stages of development follow a course similar to that in animals. In the later stages, however, strong contrasts become apparent: the body of a developing animal very soon breaks contact with its embryonic state and retains no patches of actively growing tissue corresponding to the whole mass of the infant organism. On the other hand, in the body of a plant such as a tree there remains throughout life, at the apex of the main stem, at the tips of the leafy branches and of the thousands of slender roots which slowly spread through the soil, a small patch of living substance which in structure and power of growth is practically identical with the growing substance of the embryo plant. This distinguishing feature of a plant is illustrated by the appended diagrams.

During the earlier stages of its existence the whole of the body is in a state of active growth and consists of embryonic material; later, growth tends to become localized at the ends of the elongating stem and root, and by degrees the contrast between actively growing and adult or permanent regions becomes more marked. Stems and

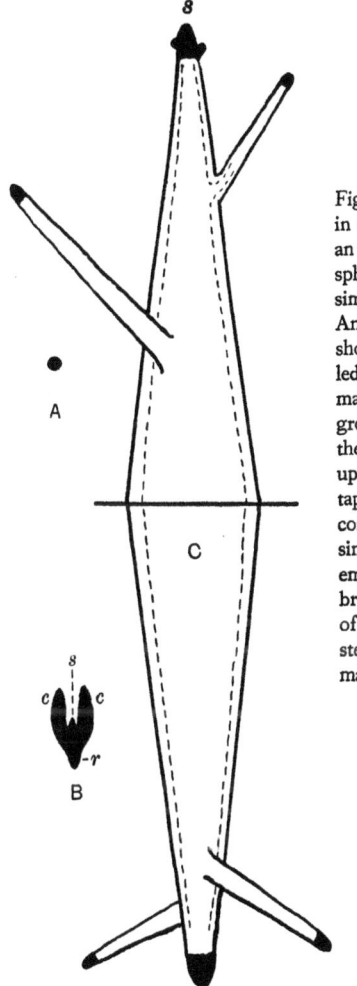

Fig. 1. Diagrams illustrating stages in the development of a tree such as an oak. *A*. A very young, more or less spherical embryo consisting of a few similar cells all actively growing. *B*. An embryo removed from a seed showing the two seed-leaves (cotyledons c, c), the future stem s, and the main root r composed of actively growing cells. *C*. A much older plant: the main stem, with branches, tapering upwards from the ground; the root tapering downwards. The black tips consist of groups of cells precisely similar to those of which the whole embryo (*A* and *B*) consists. The broken lines in *C* show the position of the cambium (page 7) within the stem and root. The horizontal line marks the ground-level.

roots produce branches, and as each branch increases in length and breadth the actively growing tissue becomes segregated at the tips. Each year the number of branches increases and there is a corresponding increase in the number of separate patches of tissue that remain permanently young. This may be expressed in another way: a very young plant is a soft growing mass, uniform in structure; as increase in size continues portions of the soft body become harder and acquire different structural features, but there always remain at the extremities of the stem, foliage-shoots and roots (Fig. 1, *C*), regions which retain the qualities characteristic of the embryo. The once continuous embryonic mass is broken up into discontinuous or discrete portions: these separate patches increase in number year by year and become more widely separated in space from the single original mass which formed the whole of the embryo.

Fig. 2. A piece of a sycamore twig from which the bark (*B*) has been partially pulled away from the wood (*W*).

This description conveys but a partial and incomplete picture of a growing tree. There are not only separated regions of permanently juvenile tissue at the tips of roots and shoots; there is also a cylinder of formative tissue enveloping the outer surface of the wood similar in structure and properties to the small body of the embryo, and to the tissue at the tips of the stem and branches. If a small branch of a sycamore be gently tapped all over with the back of a knife it is possible to remove the bark as a complete hollow cylinder and leave exposed the surface of the wood: this separation of bark from wood is the result of the rupture of the delicate cells of the intervening cylinder of formative tissue—known as the cambium—and the wetness of the exposed surfaces is due to the liquid

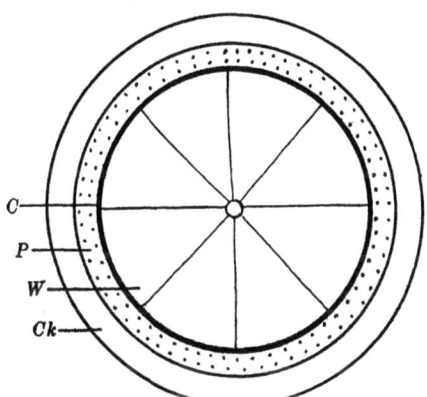

Fig. 3. Diagrammatic cross-section of the stem of a tree showing the cylinder of formative tissue (cambium, *C*) between the wood (*W*) and the bark (*P* and *Ck*). The bark includes a tissue (*P*) known as the phloem, which transports food from the leaves, and the external tissue which consists in part of cork (*Ck*).

contents of the torn cells. The cambium extends through the length of the main axis and branches and by its activity in the growing season contributes a fresh ring of wood and other additions to both stem and root (Fig. 1, *C* and Fig. 3). The trunk of an oak consists mainly of wood, the greater part of which is dead, and provides the strength necessary to support the weight of the crown and resist the force of the wind. Wood that has reached maturity, and this it does in a single season, is incapable of further growth; but as new branches are produced every year and the area of the whole tree becomes greater, additions are made to the wood through the activity of the cambium cylinder. A man may increase inconveniently in bulk long after he has reached maturity, though this is not an expression of normal and necessary development as it is in a tree.

Both animals and plants possess organs or members, such as head, arms, legs, branches, leaves and flowers. We know that our members have different forms and construction which render them well fitted to take their respective shares in the activities of life. Similarly the members of a tree differ one from another in form and structure and serve various purposes. Both plants and animals illustrate the principle of division of labour.

The living body whether of a plant or an animal is often spoken of as a machine: the several organs, like the parts of a machine, are accurately correlated and each serves special purposes. In animals there are also internal members or organs; the bony skeleton, muscles, nerves, heart and blood vessels, the stomach and many other organs. When we look at the inside of a plant we do not find any organs or well-defined structures which can easily be removed from the rest of the body. But though the body of a plant appears to be much more compact and solid than that of an animal, closer examination reveals the existence of many differently constructed parts or tissues: it does not, however, include such things as nerves, lungs, heart and blood vessels, or intestine.

CHAPTER II

Response of Plants to Stimuli

One of the attributes of living creatures is their capacity to respond to external influences: in other words they are irritable or sensitive to stimuli. It is worth while to enquire what the term stimulus implies. In ordinary language we speak of stimulating a person by word or deed or by supplying tea, alcohol or other "stimulating" drink. In a scientific sense a stimulus may be defined as an influence which produces a change in behaviour. It is often thought of as an influence which produces movement in an organism otherwise at rest; but a reaction of this kind is really a special case representing an advanced stage of organization of the mechanism connected with perception and movement, as for example the reflex movements of animals. A human being responds quickly to external stimuli: the eye perceives some imminent danger, such as a threatened blow; the nerves transmit the stimulus received by the eyes and the muscles raise a protective arm. This is the type of irritability with which we are most familiar, but it is not universal among living organisms though there are examples of it in plants as well as in animals. When the leaf of the sensitive plant (*Mimosa pudica*) is touched, certain things happen in the living cells which alter the stability of the plant-organ; the leaflets fold up in turn and eventually the whole leaf droops.

The external influence does not provide the energy

expended in the movement of response; it either modifies the direction taken by release of energy from the organism or actually sets going processes, otherwise quiescent, which result in movement and a release of energy. A light applied to gunpowder initiates chemical action which causes an explosion: there is a violent and sudden manifestation of energy and the amount represented by the explosion is measured by the quantity of powder, not by the energy of combustion of the match. The general case in which a stimulus, that is a change in the environment to which the organism is sensitive, produces a change in behaviour, may be illustrated by the following analogy. A boat is driven by a pair of independent propellers, the speed of each being controlled by the amount of pressure exerted on an electric button. By increasing the pressure on the left button we increase the speed of the left propeller, and the boat turns. We may say that the boat has been turned by the alteration of pressure, but the energy expended in the turn is in no way derived from the pressure on the button but from the generators.

Let us now see what evidence can be given in confirmation of the statement that plants respond to external stimuli. It is well known that if a plant is placed near a window, where it receives much more light on one side than on the other, the stem bends towards the source of light and the leaves tend to place their flat surfaces at right-angles to the incident rays. Light acts as a stimulus to which the plant responds. A plant has no sense-organ which even remotely resembles an eye: none the less, it seems reasonable to assume that there is some definite part which perceives the stimulus of light, some receptive

RESPONSE OF PLANTS TO STIMULI

region. By means of a simple experiment it can be demonstrated that there is a sensitive region corresponding, in the part it plays in setting in motion a response, with the eye of an animal, though it is not comparable in structure. Our eyes perceive the vibrations emitted from the sun; they are the perceptive organs and receive the stimulus of light, but the response to the stimulus received by the

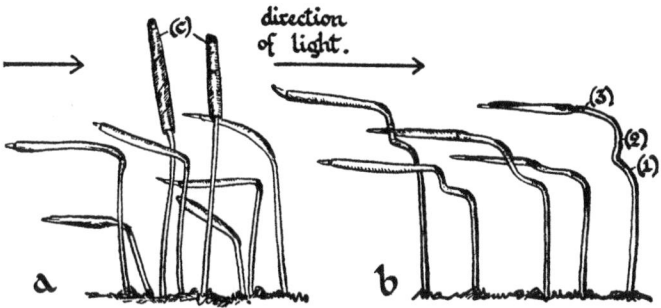

Fig. 4. Illustration of the effect of light upon seedlings. *a*. Seedlings grown in darkness; two were capped with tinfoil (*c*) and all were at first erect. They were then exposed during the day to light coming from one side: those with caps (*c*) continued to grow vertically upwards; those with their tips uncovered curved towards the light. *b*. These seedlings were left in darkness a second night. Next day they showed a double curvature: (1) is the bend caused by the light on the first day; (2) the vertical growth, due to the response to gravity, during the night; (3) the horizontal growth on the second day in response to the stimulus of light. (After Godwin.)

eyes is made by a part of our body some distance from the receptive organ. The following experiment is easy to perform:

Grow a crop of seedlings in a flower-pot, preferably seedlings of a grass (e.g. *Setaria*, Italian millet); when they are about an inch high cover the tips of some of them with caps of tinfoil in order to screen them from the light.

Place the pot in a wooden box with one side made of a sheet of glass. If the box is now put on a table where it receives light through the glass side the seedlings will be directly illuminated on one side only. After a few hours it will be noticed that the seedlings with uncovered tips are bent towards the light while those with the tips excluded from the light have continued to grow vertically upwards. The difference in behaviour of the two sets of seedlings suggests that the tip must be the sensitive region. The response to the stimulus of light on the part of the uncovered seedlings is expressed by curvature of the slender stem some distance below the tip, and the bending is due to a difference in behaviour of the two sides of the growing organ. On the side away from the light, growth is increased and simultaneously growth on the illuminated side is decreased; the total growth remains practically the same. The important point is that there is a redistribution of the total energy of growth rather than any increase or decrease. It would seem that light causes changes in the living cells which interfere with the uniform upward growth. The tip perceives the stimulus of light: how, one asks, is the effect of the stimulus transmitted to the growing region? When the stem is growing vertically the tip exercises a uniform control over the part below it. The control operates through certain substances produced in the cells of the tip which move through the living tissues below the tip to the growing region. Normally these substances are uniformly distributed and so exercise the same influence on all sides. If, however, the stem is illuminated on one side more strongly than on another, control of the apex is no longer uniform because the unequal supply of light causes an unequal distribution of the growth-regulating substances: this leads to a lop-sided control and is expressed by unequal growth and consequent bending of the stem.

We will next consider a plant's response to an entirely different stimulus, namely Gravity. A stone falls to the ground after being thrown into the air because the momentum given to it by the thrower is eventually spent and it falls in response to the attraction of the earth. It needs no botanical knowledge to appreciate the fact that the root of a seedling grows vertically down and the stem elongates in the opposite direction. The habit of stems and roots to behave in this way seems so obvious and indeed necessary that few people are curious to discover why a plant does what it is clear it ought to do. In whatever position a seed is placed in the ground the young root always grows downwards and the young stem behaves as a stem should behave. If a seed from which the first root of the embryo plant has pushed its way out is taken up and replaced with the root pointing upwards, after a short time the normal position is regained. It is all very well to say that roots grow down in order that they may fulfil their functions as roots; but there is no evidence that the plant is conscious as we are and its movements are mainly directed by external influences. What then is the stimulus which controls the movements that appear to be purposeful?

Rather more than 125 years ago an Oxford man, Thomas Knight, who was a keen naturalist and sportsman, read a paper before the Royal Society of London on the direction of growth of the young stems and roots of seedlings. He said that in the opinion of some naturalists gravity exercises a directive influence upon seedlings, but he also stated that this view was not reinforced by any experimental proof. Knight set himself to test by experiment the efficiency of gravity as a directive influence on growth. Assuming gravity to be the directive influence, he realized that it can produce the effects ascribed to it only when the seed is at rest, or rather in a fixed position relative to the attraction of the earth. If he could rapidly change the position of a germinating seed the action of gravity would be suspended; there would be no time for it to achieve any visible result. He therefore made a small wheel a foot in diameter and attached pea seedlings to its rim, arranging them so that the young roots pointed in all directions. The wheel and seedlings were enclosed in a box and driven in a vertical plane by a stream of water, which was conveniently handy in his garden, at a speed of at least 150 revolutions a minute. As growth proceeded all the roots bent outwards, taking up a position away from the centre of the wheel and parallel to the radii. The stems, on the contrary, turned towards the centre. He then placed the wheel in a horizontal plane and caused it to rotate rapidly: in that position the effect of gravity could not be suppressed because the seeds were always in the same position relative to the earth's attraction. The result was: the roots pointed downwards at an angle of about 10° below the horizontal line and the stems took up positions 10° above it.

A rapidly revolving wheel creates a powerful centrifugal force and bodies on its circumference, if free to move, fly outwards. If a bucket full of water is swung rapidly round at the end of a rope tied to the handle the water is driven by centrifugal force against the bottom and, so we are assured, is not spilt. On a vertically revolving wheel centrifugal force is superimposed on gravity. Gravity, of course, continues to exert an influence but the effect of this is neutralized by the rapid alteration in the positions of the seedlings relative to the earth. On the other hand, centrifugal force continues to act and always in the same direction, from the centre of the wheel outwards. In effect, therefore, one may say that centrifugal force has been substituted for gravity. The seedlings react to this substituted force as they react to gravity when they are growing in the ground. But when the wheel is revolved in a horizontal plane gravity continues to exert its influence, the amount of which can be estimated. If centrifugal force were the only one in action the roots and stems would respond to it in the same way whether the wheel revolved vertically or horizontally. But Knight found that on the horizontal wheel they grew respectively downwards and upwards at an angle of about 10°. This shows that centrifugal force had not entirely its own way or the roots would have grown in the plane of the wheel: gravity deflected them, though only slightly. In other words, centrifugal force caused the roots to grow in a direction 80° from the vertical, leaving 10° as the measure of response to gravity. When the horizontal wheel was revolved more slowly, gravity exerted a relatively stronger influence.

From these experiments Knight correctly concluded that the force which exerts an influence on growing seedlings is external to the plant and not within it. Gravity, as we have seen, stimulates stems to grow away from the centre of the earth and roots to grow towards it: this fact is expressed by calling roots geotropic, or positively geotropic, and stems negatively geotropic. The term tropism is used for various movements produced in plant-organs by stimuli (Gk. *tropeo*, to turn): geotropism or geotropic curvatures denote the response to gravity, the attraction of the earth (Gk. *Ge*, the earth); heliotropism the movements caused by light (Gk. *Helios*, the sun), and so on.

Since Knight's day much work has been done on the effect of gravity on plants. Charles Darwin made many experiments on movements of stems and roots. Among other questions he put to the plant was this: which part receives or perceives the stimulus of gravity? A young plant, such as a seedling of a broad bean which is large enough to be handled conveniently, is taken out of the ground when the root is two to three inches long and the seed is pinned to a piece of cork fixed in the hollowed out inside of the stopper of a wide-necked glass bottle containing a little water to keep the air moist. The seedling is so placed below the stopper that the root lies horizontally. After a short time the tip is seen to point downwards as the result of the bending of the responsive or motor part of the root (Fig. 5, *A*). Darwin repeated the experiment with a root which he had decapitated by cutting off the tip: there was no response to the stimulus of gravity and the root continued to grow horizontally. From this and other experiments he concluded that the tip is the sensitive region. Several years after Darwin and his son Francis performed these experiments it was objected that the roots had not been fairly treated: the removal of the delicate apex might well cause an injury sufficient to render the organ incapable of behaving in a normal manner. A more delicate test was devised which confirmed Darwin's conclusions. The tip of a root was made to grow into a small piece of glass tubing bent into the form of a boot; the root was placed horizontally with its imprisoned tip pointing down (Fig. 5, *B*). In this position the horizontal part of the root continued to grow forward without any curvature. So long as the tip is vertical it is, so to speak, satisfied. If, however, the boot is placed with the closed end held in a horizontal position (Fig. 5, *C*) the tip is stimulated, but though the resulting excitation is transmitted to the hinder part, no bending can now move the fixed tip. When the closed end of the boot is horizontal the hinder part of the root, which responds by unequal growth to the stimulus received, makes desperate efforts to bring its tip into the vertical position, and though there is much bending the tip remains unsatisfied.

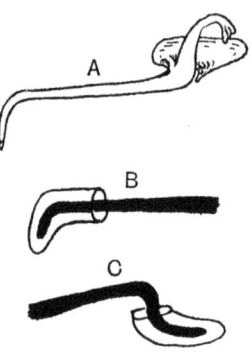

Fig. 5. The response of roots to the stimulus of gravity. *A*. The root of a bean seedling placed horizontally; the tip has taken up a vertical position. *B*. The tip of the root imprisoned in a glass boot in a vertical position. *C*. The closed end of the glass boot placed horizontally; the hinder part of the root is bending in response to the stimulus transmitted from the unsatisfied tip.

The late Sir Francis Darwin made use of the glass boot method and had a share in contributing to the confirmation of his father's views: he experimented with small seeds much lighter than broad beans. The tip of a slender seedling root was persuaded to grow into a glass boot which was fixed with its closed end in a horizontal position, the rest of the seedling with the seed being left free to move. The tip in the horizontal or unsatisfied position received the stimulus and transmitted an excitation to the hinder part which responded by bending, but as the tip could not be brought into the vertical position, messages continued to be received and the seedling occasionally tied itself into a knot (Fig. 6).

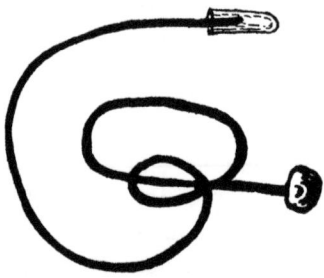

Fig. 6. A slender seedling attached to the seed with its tip held horizontally in a glass tube: the bending of the seedling is the expression of futile efforts to bring the tip into the vertical position. (After F. Darwin.)

It seems probable that this sensitiveness of the root-tip to gravity is similar to the effect already described for the sensitiveness of the stem-apex to one-sided illumination. So long as the root-tip is vertical the force of gravity is exerted equally on all sides, and the control exercised by the tip over the growing region behind it remains uniform. When the root-tip is horizontal the upper and lower sides are unequally affected by gravity, and this leads to a lop-sided control of the growing region.

Recent experiments of another kind have afforded additional confirmation of Charles Darwin's conclusions: it has been found that if the tip which has been cut off is

RESPONSE OF PLANTS TO STIMULI

replaced, the normal sensitiveness is restored and the root responds to the stimulus of gravity as it did before decapitation.

In addition to light and gravity there are many other stimuli which exercise a directive influence on plant organs: one additional illustration may be given. Roots absorb water from the soil and it is clearly advantageous that in comparatively dry ground the absorbing roots should grow most readily towards places where water is available. It has been found by observation that roots react to different amounts of moisture in the soil surrounding them; they turn towards regions of higher water-content as stems turn towards light. The joints of drainage-pipes occasionally become loose, and if there is more moisture within the pipes than in the surrounding soil the slender branches of roots insert themselves into the loose joints and by spreading uniformly may completely block the channels.

The foregoing account of geotropic curvatures deals only with the main root of a seedling, not with the subsequently formed lateral roots. If the seedling of a broad bean is examined when it has been growing several days it will be found that, though the main root is vertical, there are more slender lateral roots which grow out from it and take up an oblique and not a vertical position; they slope outwards and downwards: this shows that the side branches respond in a different degree and grow in a direction between the horizontal and vertical. When we examine the older root-system of a plant we find that the finer branches grow through the soil in many directions and apparently pay no attention to gravity. This differential behaviour of the various parts of the whole root-system affords an interesting illustration both of the correlation of response to a stimulus and of the efficiency of the plant. The ordinary explanation of the course followed by the different parts of a root-system in the soil is simply that roots are in search of water and do their best to find it. In scientific language we say that the main root of a young plant is positively geotropic, the lateral branches or secondary roots are influenced to a less extent by gravity, while the still later formed or tertiary roots appear to be insensitive to gravity. These differences in the behaviour of roots are obviously advantageous to the plant: if all the roots grew vertically down in one fibrous bundle they would tap only a comparatively small area of soil, but behaving as they do a large area is laid under contribution and the root-system is enabled to carry out its work with efficiency. Finally, it must be remembered that under natural conditions a growing plant is exposed to many stimuli simultaneously, e.g. gravity, water, light, etc.; and the movements made by the various organs represent responses to more than a single stimulus.

We have defined a stimulus as an influence producing a change in behaviour, and the examples chosen illustrate the effects of influences which are usually called external stimuli because they are outside the plant though in its immediate environment. There are, however, many changes in the behaviour of plants due to influences within its body; these are known as internal stimuli. We may think of a plant as sensitive to the particular kind of environment or field in which it finds itself at any one time; it is exposed to certain conditions both external to it and within it. A familiar example of an internal influence or stimulus will make this distinction clearer. It is not uncommon to see a tree, such as a fir, in which the leader —that is the main stem—has been killed above the topmost branches: when this occurs one of the branches, which normally grow more horizontally than vertically, bends upwards and eventually takes the place of the leader. The results of research lead us to believe that the main stem when it is healthy exerts a certain influence on the direction of growth of the lateral branches; it has an inhibitory effect upon them which prevents their upward growth and modifies the way in which they respond to the gravity-stimulus. On the death of the leader this inhibitory influence is no longer operative and a branch then takes over the rôle of the main stem. This effect is no doubt produced by some substance made in the living cells of the apex and conveyed to regions below the apex where it exercises a controlling or inhibitory influence on the lateral branches.

CHAPTER III

The Superiority of Green Plants to Animals

We will now discuss the statement made at the beginning of the first chapter that in one important respect plants, or at least green plants, are superior to animals. A machine is useless without some driving force: a boiler generates heat derived from the burning of fuel; a dynamo producing an electric current may be driven by a waterfall or a heat-engine. Plants and animals for our present purpose may be regarded as machines made up of various parts differently constructed and accurately fitted together: they are, however, distinguished from all other machines by the possession of protoplasm. A machine in motion and a living organism use energy; they do not create new energy, they obtain energy from some external source. Green plants derive their energy ultimately from the sun as light; animals, on the other hand, obtain their energy from the food which they eat. Animals eat organic food and from it derive the energy which they use in movement and in maintaining the heat of their bodies: plants build up organic substances and store energy. This difference is fundamental and will be considered at some length. The transactions of the material universe are conducted, as the famous Cambridge physicist, Clerk Maxwell, said rather more than fifty years ago, on a system of credit: each transaction consists in the transference of so much energy from one body to another. This act of transference is called work.

Life depends upon incessant changes within the living cells; changes which do not create energy but by the rearrangement of material release energy and convert one kind of energy into another. New energy is never created and energy is never permanently lost: when a brake is applied to a revolving wheel work is done and the energy of motion of the wheel is diminished and apparently, though not actually, destroyed because the pressure of the brake on the rim of the wheel produces heat. Energy of motion is transformed into heat energy. The phrase "conservation of energy" expresses the fact that energy is not produced *ab initio* and is never lost: the sum total remains the same. Another aspect of this concept of energy must be made clear before we proceed further, namely the difference between what may be called energy of position and energy of motion, or in more scientific language potential energy and kinetic energy (Gk. *kineo*, to move). In throwing a stone upwards we expend energy and impart to the stone energy of motion: as the stone goes higher this kinetic energy is gradually expended until eventually there is none left. But there has been no absolute loss: the kinetic energy has been transformed into potential energy or energy of position. A bent spring represents a store of energy: in winding up a clock we expend energy which adds to the amount of potential or stored energy in the mechanism of the clock.

The next step leads to an examination of the statement that food is a source of energy. This question necessitates a short digression into the realm of Chemistry, the science which deals with the structure of all homogeneous kinds of matter, with the changes such matter undergoes and with the nature of the substances produced. It may be thought unnecessary and tiresome to introduce even a small amount of Chemistry into a discussion on the nature of plants. Plants, like all other things in the world, are made of substances with which chemists are concerned and it is clearly impossible to appreciate the mechanism of life unless we understand some of the terms relating to the constitution of matter. Earth, air, fire and water were formerly and indeed are still sometimes called the four elements of the material world. Chemists apply the term element to all the simple, chemically indivisible substances of which matter consists. In recent years it has been proved that certain radioactive elements spontaneously produce matter unlike themselves; but these are elements as chemically defined. The number of known elements is about ninety: some exist naturally as gases, e.g. oxygen, hydrogen, nitrogen; some as metals, e.g. gold, platinum, copper; and many others occur in nature only in combination with other elements.

Every element consists of particles known as atoms (from the Greek *a*, not, and *temno*, to cut), millions of times smaller than the smallest fragments which can be seen under the highest magnifying power. It is customary in chemical language to speak of an atom as the smallest particle of an element which can exist in combination with another atom. Atoms of all the elements have different weights. We now know that even an atom is not an indivisible entity: it is an inconceivably minute microcosm having a definite structure. In the centre is a nucleus charged with positive electricity and arranged around the nucleus, but at some appreciable distance from it, are one or more electrons—units of negative electricity. The atom of hydrogen, the simplest of all, is about half a hundred-millionth of an inch in diameter and consists of a nucleus or proton associated with a single electron. Atoms have never been completely broken up in the laboratory; but it is possible to chip pieces off them. Some of the heaviest atoms, e.g. the atom of radium, undergo spontaneous disintegration and break up into smaller atoms having a different structure emitting in the process the well known X-rays.

Another term, molecule (diminutive of the Latin *moles*, a mass) is applied by chemists to the smallest particles which exist as separate entities, not necessarily combined with other molecules. Compounds consist of at least two elements held together by chemical affinity and in definite proportions.

A description of a relatively simple compound will serve to illustrate some of the fundamental principles of the constitution of matter which it is important to understand. Water is a chemical compound consisting of one atom of oxygen closely united by electrical or magnetic forces with two atoms of hydrogen (water producer: Gk. *hudor*, water; *genea*, birth) and is represented by the formula H_2O. If hydrogen is burnt in oxygen the molecules of the two gases fly together and by their union lose their individuality and form an entirely different substance, water. Their combination is accompanied by the liberation of energy in the form of heat. The energy stored in a single molecule of water is less than that represented by the hydrogen and oxygen molecules before they come together. The pulling to pieces of water requires expenditure of energy equal in amount to the energy liberated when hydrogen and oxygen combine. If an electric current be passed through water the energy so supplied overcomes the force with which the two elements are held together in the compound. The gas oxygen is a particularly active element; it readily combines with other elements producing numerous compounds known as oxides. Water is an oxide of hydrogen formed when hydrogen undergoes combustion, or burns, in oxygen.

We will next take an oxide which occupies a unique position, namely the oxide of carbon called carbon dioxide because it consists of one atom of carbon combined with two atoms of oxygen. When carbon, e.g. charcoal, is burnt in oxygen the gas carbon dioxide (CO_2) is formed with the evolution of heat. The energy stored in the CO_2 is less than the amount stored in the separate molecules of the two elements.

When coal is burnt a large amount of energy is released as heat and one of the products is carbon dioxide gas: it is the energy set free in combustion which is used as a source of power. At this point it may be helpful to refer briefly to a most important part of the evidence in support of the statement that plants are superior to animals. A fuller discussion with some accounts of experiments is given on a later page. Coal consists largely of the altered remains of plants which, we assume from our knowledge of living plants, contained in their bodies and framework many complex compounds derived in the first instance through the agency of the green leaves from carbon dioxide

OF GREEN PLANTS TO ANIMALS

gas in the atmosphere of the Coal Age. The atmosphere in that remote period of geological history, separated from the present by possibly about two hundred million years or more, when enormous areas were covered with forests, supplied the material of which most of our coal is made. The composition of the atmosphere in the Coal Age was probably much the same as that of the air which envelopes the world to-day. Air is a mixture of gases: it contains by volume 78·03 per cent. of nitrogen; 20·99 per cent. of oxygen; 0·03 per cent. of carbon dioxide: the remaining 0·95 per cent. is made up of small amounts of certain inert gases, argon, neon, helium, krypton, xenon and a very slight trace of hydrogen, which need not be further considered. Nitrogen is so called because it is one of the elements in nitre (saltpetre); by the French it was called Azote (Gk. *a*, not; *zoe*, life), a term descriptive of its inability to support life. Oxygen, which forms about one-fifth of the air, readily combines with other elements and is a good supporter of combustion, that is to say, substances readily burn in oxygen. If a burning match be gently blown out and the still glowing splinter plunged into oxygen it bursts into flame.

The carbon dioxide, present in the proportion of three parts of gas to 10,000 of air, must be more fully described. We assume that the proportion of carbon dioxide in the Coal Age atmosphere was approximately the same as at present: it has often been suggested that the amount was greater, but that cannot, of course, be proved. Animals make no use of this gas. With green plants the situation is entirely different: the carbon dioxide gas in the air is the source from which they obtain the whole of the stock of carbon used by them in building up carbon compounds, e.g. sugars, starch, and other more complex substances necessary for our existence and indeed for the whole animal kingdom. It is difficult to believe that a plant can obtain from the air the substance which we are familiar with as charcoal, lamp black, graphite and diamonds. Let us then see how it is possible to prove the presence of carbon in the air. If a piece of charcoal be heated red hot and plunged into a jar of oxygen gas it burns brightly and in doing so combines with the oxygen to form an oxide, carbon dioxide: this oxide is invisible but its presence is easily demonstrated by adding lime water—that is slaked lime and water—and shaking the bottle in which the charcoal was burnt: the lime water becomes milky and this is evidence of the presence of carbon dioxide. The turbidity is due to the formation of particles of calcium carbonate (chalk) from the combination of the carbon dioxide with the lime water. To prove that carbon can be extracted from carbon dioxide, introduce burning magnesium wire into a jar of the gas: it continues to burn, though not nearly as brightly as it does in the air (e.g. when photographs are taken by magnesium light), and soot (carbon) is deposited. The intense heat has supplied the energy necessary to split up the carbon dioxide into carbon and oxygen.

In order to break up CO_2 in the laboratory a temperature of about 1500° C. is required, a temperature fifteen times as high as the boiling-point of water. This shows that the two elements carbon and oxygen are held together by very powerful forces which can be overcome only by the expenditure of an enormous amount of energy.

It is clear therefore that at the temperature of the body an animal cannot utilize carbon dioxide as a source of carbon. Green plants have the means of providing the requisite energy in the substance chlorophyll (Gk. *chloros*, grass green; *phyllon*, a leaf), which gives them their colour because it reflects the green rays of light and absorbs others. Chlorophyll absorbs certain rays of light, as will be demonstrated later, and thus provides radiant energy by which the dissociation of CO_2 is effected and the carbon rendered available for the manufacture of sugars, starch and other compounds. We must now give closer attention to the element carbon: its atoms are relatively complex and have in a remarkable degree the power of entering into combination with the elements oxygen and hydrogen with which they form hundreds of carbon compounds. These are often spoken of as organic compounds because it was formerly believed that they could not be produced artificially but only through the agency of some mysterious vital force. The name carbohydrate is given to many of these compounds because in them oxygen and hydrogen occur in the same proportion as in water (H_2O). Carbohydrates are among the most important constituents of food used by animals and plants.

In 1828 a German chemist, Friedrich Wöhler, wrote a paper in which he said that his research had led to an unexpected result: he found that by the combination of cyanic acid with ammonia urea is formed, "a fact that is noteworthy since it furnishes an example of the artificial production of an organic, indeed a so-called animal substance, from inorganic materials". Urea is not a carbohydrate; it is more complex in that it contains in addition to carbon, hydrogen, and oxygen, the element nitrogen: it is one of the waste products of animals and occurs in urine. Since the dividing line between what are still called organic and inorganic chemistry was crossed by Wöhler's important researches, many other organic compounds have been produced in the laboratory, but by no means all: the plant remains the only laboratory where the carbohydrates and other carbon compounds are manufactured at ordinary temperatures.

Carbohydrates, as already stated, are essential for animal life: they serve as fuel and as a source of energy are comparable with coal in a boiler. We speak of food as a source of energy: the energy contained in carbohydrates must be released in order to be of use in driving the living machinery.

We have seen that in the combustion of coal, carbon dioxide is one of the products and with it kinetic energy. Similarly the carbohydrates of living organisms constantly undergo combustion in the presence of oxygen, carbon dioxide being evolved and energy released. Carbohydrates contain a much greater store of energy than occurs in carbon dioxide: it is during the conversion of substances richer in energy (carbohydrates) into a gas with a lower energy content (carbon dioxide) that energy is released and rendered available to the plant. If the carbohydrates were burnt as coal is burnt the heat produced in this sudden change would be so much energy lost to the plant; but what actually happens is a gradual transformation, in stages, of the carbohydrate to carbon dioxide: part of the energy released in the course of this transformation supplies the power required for growth movements and other activities of the plant.

It is true of both animals and plants that they obtain energy from the combustion of carbohydrates; they obtain energy also from the combustion of the nitrogen-containing compounds known as proteins, but these need not be considered at the moment. The outstanding fact is that the carbohydrate fuel is made in the first instance by green plants, the carbon used for this purpose being extracted from the carbon dioxide in the air through the agency of the chlorophyll in the living cells. Thus we see that animals are inferior to green plants in that they cannot lay the foundation stones of carbohydrates and proteins.

CHAPTER IV

How a Plant obtains Energy

Much has been said about energy, but with no reference to the possibility of measuring it. Nowadays it is not uncommon to meet people who like to talk about calories; they offer advice at the dinner-table on the selection of food which is richest in calories. The term calorie is used for the unit of heat required to raise the temperature of 1 grm. of water through 1° C.; another unit, known as the British Thermal Unit, is the amount of heat required to raise the temperature of 1 lb. of water through 1° F., and this is equivalent to about 252 calories. By measuring the change in temperature of water enclosed in a special apparatus (calorimeter) it is possible to measure the amount of energy liberated, or absorbed, as heat during chemical combinations or dissociations of compounds. It has been found, for example, that when 12 grms. of carbon are burnt and the carbon unites with the oxygen of the air to form carbon dioxide, 97,000 calories are set free. In estimating larger quantities of energy, as for example in experiments on human beings, it is usual to employ a different unit, the kilogram calorie (sometimes written Calorie), which is 1000 times as large as the ordinary calorie. It has been found that a man engaged in sedentary work needs about 3000 kilogram calories a day; when he is employed on hard physical work 7000 kilogram

HOW A PLANT OBTAINS ENERGY

calories are required. This simply means that the amount of fuel necessary, whether carbohydrates or other sources of potential energy, varies with the quality and amount of the work being done. It is well known that human beings like other living organisms give off carbon dioxide in respiration. Oxygen taken in by the lungs is distributed by the blood-stream to the various organs of the body; it enters into combination with the complex substances which serve as food and the carbon dioxide given off is one of the products of the breaking up of these substances. In plants there is no special breathing organ: as in the animal so also in the plant every living cell respires. Animal cells obtain oxygen from the blood and give off carbon dioxide to the blood; plant cells take oxygen from the air which circulates freely both outside and inside the plant; they give off carbon dioxide to the air in the spaces which form an aerating system within the plant-body, and take oxygen from the internal air.

We will now consider more fully the nature and source of the materials of which the food of green plants consists. Wheat and other cereals, cabbages and other vegetables supply part of the fuel from which we derive calories. We know that a plant draws much of the raw material which is eventually used in the construction of sugars, proteins and other substances, from the ground. The carbon is obtained from the air. Before enquiring into the nature and source of the material obtained from the soil, let us see what substances occur in a grain of wheat and a lettuce leaf. A grain of wheat contains 13·65 per cent. of water, 84·5 per cent. of combustible substance and 1·8 per cent. of ash which is left on burning the grain. A lettuce leaf contains about 94 per cent. water, 4·6 per cent. combustible substance and 1 per cent. ash. If a plant is dried by exposing it to a temperature a little above 100° C. (boiling-point) the water is driven off: the dry material left consists of 45 per cent. carbon, 42 per cent. oxygen, 6·5 per cent. hydrogen, 1·5 per cent. nitrogen, and 5 per cent. mineral constituents which form the ash. Burning drives off certain gases, and these on analysis are found to include oxygen, hydrogen, nitrogen and the element carbon, all of which are constituents of living plants and animals. Analysis of the incombustible ash reveals the presence of sulphur, phosphorus, chlorine and silicon which in combination with the metals potassium, magnesium, calcium and iron (elements which are also present in the ash) form salts. A salt is a compound formed from an acid by the substitution of a metal for the element hydrogen. Carbonic acid, made by dissolving carbon dioxide in water, as in effervescing mineral waters, contains carbon, oxygen, and hydrogen (H_2CO_3): carbonate of lime (chalk; $CaCO_3$) is a salt of carbonic acid and consists of carbon, oxygen and the metal calcium. There are about seventy elements present in soil but only a few are invariably found in plants: these are hydrogen, oxygen, chlorine, sulphur, nitrogen, phosphorus, silicon, carbon, potassium, sodium, calcium, magnesium and iron. Of these, nine are ordinarily found in the ash of plants: chlorine, sulphur, phosphorus, silicon and the metals potassium, sodium, calcium, magnesium, and iron. These elements are derived mainly from minerals in rocks which are constantly undergoing decomposition and chemical change: silicon is very abundant as the essential element

HOW A PLANT OBTAINS ENERGY

in sandstones; potassium, sodium and calcium occur in such minerals as those of granite and other rocks; calcium is abundant also in limestones which consist mainly of calcium carbonate. The elements named occur as salts such as sulphates, phosphates, nitrates, carbonates, chlorides derived respectively from sulphuric acid, phosphoric acid, nitric acid, carbonic acid and hydrochloric acid.

It is easy, by slow heating so as to cause charring, to show that the body of a plant contains a large amount of carbon. It is also easy to demonstrate that the carbon, which forms about 40 per cent. of the dry weight of a plant, is not obtained by plants from the soil: a plant can be grown perfectly well in a vessel containing water to which have been added various salts but without any trace of carbon. In order to discover the nature of the material with which a green plant must be supplied from the soil the first step is to analyse the plant-body: having ascertained which elements are usually present we then add small quantities of salts containing them to distilled water in a glass vessel. A plant is fitted into a hole cut through a cork in the broad mouth of the vessel with the roots immersed in water. The glass vessel should be covered with black paper so that the roots may grow under conditions which are unlikely to encourage the growth of foreign organisms; the water should be kept aerated by blowing in air, and from time to time the water and salts should be renewed. Experiments have proved that when a vigorous seedling, say of a broad bean, is placed with its root in a vessel containing about one litre of water to which the following substances have been added in the proportions named, it develops normally and eventually produces flowers:

Potassium nitrate	1 grm.
Ferrous phosphate	0·5 grm.
Calcium sulphate	0·25 grm.
Magnesium sulphate	0·25 grm.

Certain other elements are essential, e.g. boron, zinc and manganese, but the very small amounts of these elements which are needed are usually available as impurities in the other salts. Potassium nitrate is better known as saltpetre: it consists of the elements potassium, nitrogen and oxygen and its formula is KNO_3. Ferrous phosphate contains phosphorus, an element which does not occur uncombined in nature, iron, and oxygen; it is a constituent of many rocks. Calcium sulphate consists of calcium, sulphur, and oxygen, and in certain forms is known as gypsum and alabaster ($CaSO_4$). Magnesium sulphate consists of magnesium, sulphur, and oxygen and is familiar as Epsom Salts.

The several elements included in these substances play various parts in the life of a plant: the element magnesium, for example, is one of the constituents of the green colouring matter, chlorophyll, and that all-important substance is not formed unless the plant has access, as it almost always has in the soil, to iron. If, for example, a culture-mixture with no trace of iron in it is given to a plant, the leaves are yellow and not green. The nitrogen, obtained from the potassium nitrate in the above mixture, is an essential constituent of the proteins, those complex substances which with carbohydrates are essential for the

HOW A PLANT OBTAINS ENERGY

nutrition of animals. Phosphorus and sulphur are also necessary for the production of proteins and the living protoplasm. Though nitrogen forms 78 per cent. of air, green plants do not obtain it from that plentiful store. Unless there is a salt containing nitrogen in the culture-mixture the plant dies of nitrogen starvation. We shall revert later to the important question of the supply of nitrogen from the soil and endeavour to explain what is meant by the expression "the Nitrogen-cycle" (Fig. 21).

A word may be added on the importance of water. Roots must be in contact with water containing substances in solution; it is only substances dissolved in water that can be absorbed. Water plays an essential part in the living protoplasm of plants; it also provides the elements hydrogen and oxygen. It is only in the presence of water that the plant machinery is able to work; it is the medium in which the innumerable chemical reactions on which life depends take place. Moreover, water possesses another very important quality: it has a high specific heat, that is to say, more heat is required to raise the temperature of water through one degree than to raise the temperature of any other substance to the same amount. One advantage of this property is that changes in temperature due to heat production by the living cells or to radiant heat from outside are much smaller than they otherwise would be.

Before passing to another aspect of plant-life we may briefly summarize some of the facts so far recorded. A plant is a sensitive organism; it possesses the quality known as irritability, one of the essential properties of living protoplasm. It is able to complete its development if supplied with a few salts, water, and air. From the elements available in ordinary ground-water and the carbon in the air it builds up its body. By reason of its unique power of utilizing radiant energy from the sun it obtains carbon for the manufacture of carbohydrates and proteins. Some carbohydrates can be made artificially though not as the plant makes them; no proteins have so far been made by man. The green plant, in short, controls the food-supply and the life of the whole animal kingdom.

CHAPTER V

Cells and Tissues

Our next purpose is to obtain a general idea of the structure of a plant's body. That a plant such as a wallflower or a tree has definite members or organs, stem, root, leaf and flower, is obvious; but in order to understand the inner working of the living mechanism as a whole we must know something of the minute structure as well as the grosser features of the several parts. If a stem of some

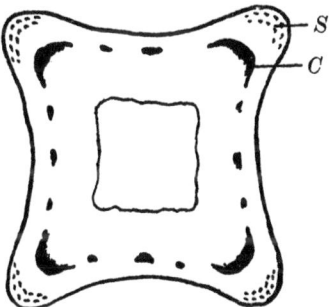

Fig. 7. Diagrammatic cross-section of the stem of the white dead-nettle showing groups of conducting tissue (*C*), groups of thick-walled cells which give the stem strength (*S*), and in the centre a hollow pith.

herbaceous plant, that is a plant which is comparatively soft and not woody, such as a white dead-nettle, which is almost always procurable, be placed with its cut end in water to which a little red ink has been added, after about half an hour we shall find, on cutting the stem across some few inches above the water, that there are well-defined red spots regularly arranged in a circle on the cut surface. The spots of colour, or a more or less continuous ring of red ink concentric with the outer surface and a little within it, show that the ink has risen along well-defined paths (Fig. 7, C). If the cut end of the stalk of a snowdrop is treated in the same way the red ink will soon be visible as delicate streaks on the white ground of the flower. It would seem, therefore, that some parts of a plant are concerned with conduction; that there is a division of labour in the material framework. If we take a leaf of a plantain (frontispiece), a plant with which most people are familiar because of its partiality to lawns, and break the blade across, it will be noticed that the two severed parts hang together by a few slender threads which are still intact (Fig. 8, *g*). These strands are in structure and formation the same as those located by the red stains in the cross-section of the dead-nettle stem. On looking through a microscope at a thin section cut from the dead-nettle stem or from the plantain leaf we see that each strand consists mainly of several very narrow tubes or miniature pipes which, as can be proved by cutting a longitudinal (vertical) section, are much longer than broad and well adapted to serve as channels for the transport of water. The whole stem is made up of microscopical units called

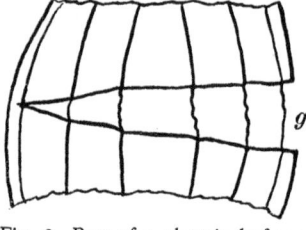

Fig. 8. Part of a plantain leaf torn across showing the veins as slender threads stretched across the gap (*g*). For a complete leaf, see the frontispiece.

CELLS AND TISSUES

cells which are not all alike and are arranged in groups, each group consisting of cells having distinctive structural features and responsible for certain definite shares in the life-processes of the organism.

It would carry us too far to pursue our examination of the internal structure of stems: the important point is that the body of a plant is made of cells and these form groups more or less clearly defined and recognizable under a low magnifying power: the groups are called tissues. A tissue may be composed of living cells, of cells that are dead, or of both dead and living cells. There are conducting tissues made of narrow and very much elongated cells which are concerned with conveying water from the roots and distributing food made in the leaves. Another kind is known as mechanical tissue: this consists of cells with strong walls and serves to give strength and support; in a mature plant the cells of this tissue are usually dead, the protoplasm having been used up in the formation of the thick walls. There are many other kinds of tissues illustrating the dependence of one part of the body upon another and the correlation of structural features with diverse functions. The cell is often spoken of as the unit of structure, and a misleading comparison is sometimes made of a plant's body with a brick wall, the bricks corresponding to cells. Such comparison is misleading for two reasons: first, the cells of a tissue are not piled one on top of another but are formed during the development of the plant by the increase in volume and subsequent bipartition of the original cells. Secondly, each cell in a tissue is not a wholly independent unit like a brick in a wall; the living contents of adjacent cells are connected through the walls of the cells by exceedingly delicate threads of protoplasm. We may think of a living plant rather as a mass of protoplasm almost completely divided by innumerable partitions into very small compartments or cells, if we remember that as the plant grows many of the cells lose their protoplasm and though dead still play an essential part in the life of the whole.

The name cell was first applied to plant structures by an Englishman, Robert Hooke, in the latter half of the seventeenth century: he examined under a microscope of his own make, which he wished to test, a thin section of cork and discovered that it was built up of a number of regular compartments which he called cells, each limited by a well-defined membrane or cell-wall. The cells of cork when fully developed on the stem of a tree are dead; their protoplasm gradually disappears as the walls become converted into the impermeable cork. Usually the term cell is applied not merely to the framework but to the living contents as well. The smallest cells visible under a very high magnifying power are those of bacteria, organisms to which further reference is made in Chapter IX; examples of cells visible to the naked eye are readily seen in the orange as little bags full of sweet fluid. An egg of a fowl is also a single cell and the largest cell is probably the egg of an ostrich.

We shall see later that there are many organisms which consist of single cells. The appearance of such an organism as the first living cell, possibly in the waters of a primeval sea, appeals to our imagination as nothing else does: we think of some creative act, some play of physical and chemical forces transforming a lifeless world into a world pregnant with the germs of an age-long succession of living creatures, at first mere specks of life which, by degrees, as evolution proceeded along different lines, gave birth to plants and animals.

Cells play many parts and together make up a heterogeneous complex which is the body of the plant. A cell of

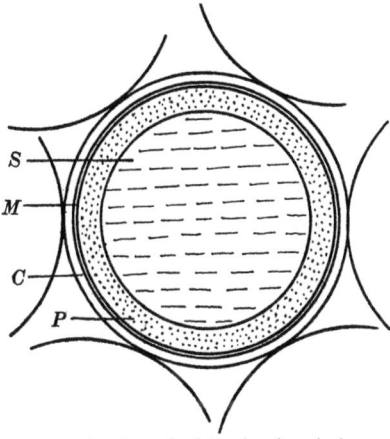

Fig. 9. A diagrammatic drawing of a fully developed plant-cell. *C*, cell-wall; *P*, protoplasm with a limiting membrane (*M*) on the outside; *S*, cell-sap.

average size is about $\frac{1}{20}$ mm. in diameter: if free to grow equally in all directions its shape is spherical (Fig. 9), but in plants the cells are for the most part contiguous and grouped into tissues in which their shape is usually polygonal as the result of mutual pressure. The greater part of the adult cell shown diagrammatically in Fig. 9 is full of watery cell-sap (*S*) separated from the cell-wall (*C*) by a layer of living protoplasm (*P*) which is bounded on its outer surface by a limiting membrane

(*M*). A typical living cell when young, as for example in an embryo or at the tip of a branch, stem or root, is filled with protoplasm and somewhere in the jelly-like or fluid protoplasm may be seen a darker and more compact mass, known as the nucleus, which is an essential part of every cell (Fig. 10, *A*). The nucleus seems to exercise a directive influence on the activities of the cell: it is a specialized portion of the protoplasm differing in structure and to some extent in composition from the rest of the contents. There is good reason for regarding the nucleus as the carrier of hereditary characters, but a fuller consideration of its dominant share in reproduction would carry us too far from our immediate purpose. Surrounding and enclosing the protoplasm is a well-defined membrane, the cell-wall, consisting of a carbohydrate called cellulose closely akin to starch in composition: its formula is $x(C_6H_{10}O_5)$, the x standing for the number of units which make up its complex molecule. Cellulose is of great economic importance: it is the substance of which cotton—the hairs on the seed of the cotton plant—consists, and it is used in the production of artificial silk and for many other purposes. The cell-wall is elastic and readily permeable by water: the egg of a plant is naked, but after fertilization it is covered with a protective wall

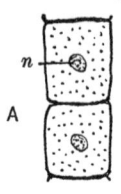
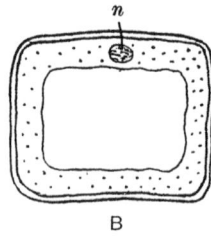

Fig. 10. *A*, two young cells full of protoplasm; *B*, an older cell with cell-sap. *n*, nucleus.

formed by the living protoplasm as a thin skin which soon increases in thickness by the introduction of additional molecules. As the cell grows the wall stretches and the whole cell may be compared with an inflated rubber ball. The cell contents increase in amount and therefore exert a greater pressure on the elastic wall and cause it to stretch. Within the cell changes occur during growth: as the volume becomes greater the protoplasm no longer fills the interior and in it are gradually formed bubble-like spaces, or vacuoles, which are filled with water holding various substances in solution. Eventually in a fully grown cell the protoplasm is reduced to a thin layer, with the nucleus embedded in it (Fig. 10, *B*); this layer is pressed closely against the wall, and the rest of the cell, as the separate bubbles, or more correctly vacuoles, coalesce, is occupied by a watery sap.

Some cells remain thin-walled throughout life, while in others the thickness of the wall is greatly increased, particularly in cells which are destined to form strengthening tissue. The conducting tissue to which reference has been made contains pipe-like channels of two kinds: the first is formed of rows of short cells arranged in regular vertical series (Fig. 11, *A*) which by the absorption of most of the cross-walls are converted into tubes (Fig. 11, *B* and Fig. 12, *V*) which serve for the conduction of water.

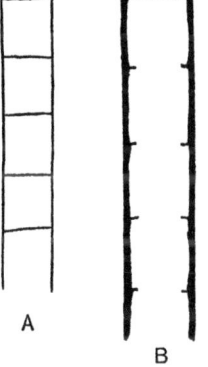

Fig. 11. A vertical row of cells (*A*) from which a vessel (*B*) has been formed by the absorption of most of the cross-walls and by the thickening of the longitudinal walls.

46 *CELLS AND TISSUES*

A cross-wall which has been retained is shown in Fig. 12, V. These walls undergo a chemical change,

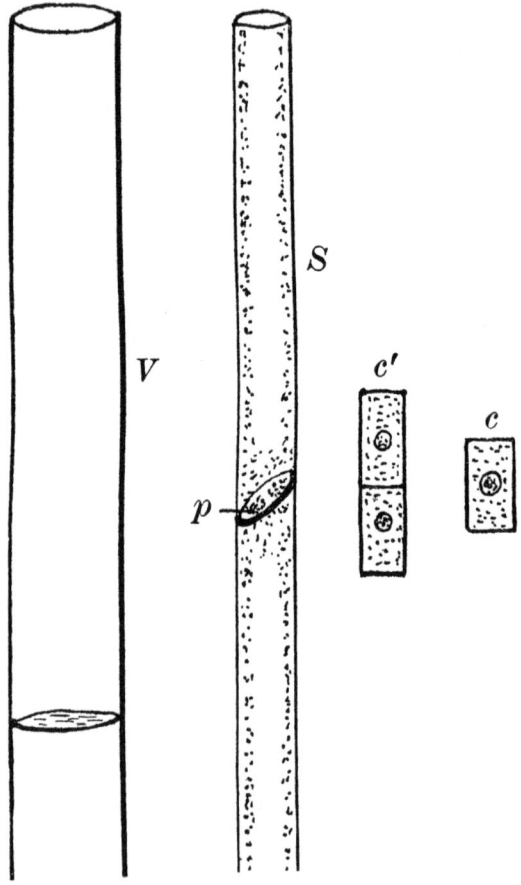

Fig. 12. A vessel (V) and a sieve-tube (S) with a sloping transverse wall perforated by pores (p). Both the vessel and sieve-tube were originally short cells full of protoplasm (c') derived by division into two from a single cell (c).

the cellulose being impregnated with a substance called lignin or wood (Latin, *lignum*, wood): these woody tubes, when mature, contain no living substance and serve merely as conductors of water. It is these tubes which make up the greater part of wood. A second type of tube in the conducting strands is known as a sieve-tube (Fig. 12, *S*): this is a more delicate and a narrower tube; the cross-walls which occur relatively far apart are perforated by groups of minute pores (*p*). In these sieve-tubes the walls are made of cellulose and a little protoplasm is retained, whereas the larger wooden tubes have walls of wood. The sieve-tubes do not conduct water but mainly organic substances, which are carried to all parts of the plant where they are needed as building material for the production of new tissue. The sieve-tubes, therefore, are the main channels through which the manufactured food is transported. Without going into further detail we may describe the conducting tissue as consisting for the most part of two sets of tubes, some relatively hard, woody, and dead which are chiefly concerned in the conveyance of water—the sap of the tree—from the ground to all parts of the plant; the others—the sieve-tubes—softer and not entirely devoid of living contents, are the delicate pipes responsible for conducting proteins and some carbohydrates from the green leaves to the tips of branches and roots and to all other regions where cell-formation is active. The green leaves may be called the factories where carbohydrates and proteins are made. It would take us much beyond the limits of this general survey of the main facts of structure and function if the various tissues were enumerated and even briefly described. For a fuller account of plant-anatomy the reader should consult botanical text-books (see p. 131).

Let us now see if we can form an idea of the number of cells in a tree: this it may at once be said is impossible; but the following statement made by a Belgian botanist is worth quoting. He pictures a forest of trees which has been growing since the beginning of the Christian era. A man is supposed to count cells all day and all night; he counts 200 a minute or 105,120,000 in a year. In round numbers he is able to count 200,000,000,000 cells in nineteen centuries. Taking the average size of a cell as 100μ, that is $\frac{1}{10}$ mm., and estimating the number of cells per square millimetre at 100 or 10,000 per square centimetre, he would have counted the cells up to a height of 20 cm., that is less than a foot, in one tree of the forest after his strenuous labour extending over 1900 years.

CHAPTER VI

The Green Leaf

Now that we have a general idea of the plan on which the body of a plant is constructed it is easier to appreciate the structural features of leaves, particularly in their relation to the important part they take in the life of a plant. It is hardly an exaggeration to describe the green leaf as one of the most wonderful things in the world: the work it does provides the store of carbohydrates, such as sugar and the more complex nitrogen-containing proteins on which practically the whole organic world subsists.

Let us first consider the more obvious and easily seen features of leaves. One of the first observers to recognize the fact that leaves have a regular and orderly disposition on stems and branches was Leonardo da Vinci, one of the greatest men of all time. If we look at the branch of a tree from above and from the direction in which light is received, we notice that there is little overlapping; the flat surfaces are for the most part freely exposed to the sun. In some plants the leaves are attached in opposite pairs, each pair at right-angles to those above and below it, while in most plants they are borne in spirals, each plant having its particular kind of spiral arrangement. The great majority of leaves are flat; their form suggests that exposure to light and air of as large a surface as possible is a distinctive and important character. There are, of course, many exceptions, as the needles of a Scotch fir and the leaves of several other trees; but the typical leaf is a thin, flat organ. Looking at the surface of a leaf in a strong light we notice that there is usually a central rib giving off lateral branches which are further subdivided into still more slender threads penetrating through almost the whole of the leaf substance. To obtain a true picture of the veins it is worth while to look in the litter of a wood-side ditch for leaf skeletons—leaves that have partially decayed and have lost the softer tissues—or to obtain skeletons by allowing leaves to rot in water. The veins serve a double purpose: the stronger ones form the main ribs of a supporting framework, and the whole ramified system acts as an efficient irrigating mechanism by which the water that has travelled from the soil through the roots and stem is distributed through the green tissue. Another service rendered by veins, in addition to their work as conductors of the rising sap, is the transport of material which the leaves manufacture to all regions where it is being used. The substances made in the living green cells of the leaf provide the material from which new cells are constructed at the growing tip of each root or rootlet, of each branch, and in the cells of the cambium, that cylinder of formative tissue mentioned in an earlier chapter. Moreover, as we have seen, the food elaborated in the leaves, carbohydrates and other carbon compounds, furnishes a store of potential energy which is utilized for the working of the plant machine.

In the account of the sensitiveness of plants to light it was pointed out that leaves show a definite response to the stimulus of radiant energy: when a stem bends towards a window it places itself approximately parallel to the

THE GREEN LEAF

incident rays of light; the leaves respond differently, they take up a position at right-angles to the light: their stalks twist and bring the flat surfaces into a position most favourable for receiving an adequate supply of sunlight.

It is abundantly clear that it must be advantageous to a plant to expose to sunlight as much leaf-surface as possible. An elm tree of average size is said to bear 7,000,000 leaves and the total leaf-area exposed to the sun is estimated at five acres. The next step is to enquire into the relation between sunlight and leaf-surface, but before doing this we must first look into the interior of a leaf and examine the structure. In most leaves there is a noticeable difference in the depth of colour on the two faces; the upper is darker than the under surface, and this is an expression of a structural difference between the two halves of a leaf as we look at a section cut at right-angles to the surface (Fig. 13, *A*; Fig. 13, *B* shows the structure of the leaf on a larger scale). We can often, by careful manipulation with a sharp knife, detach a superficial layer or skin from a leaf: after mounting a piece of it in water with the upper side uppermost on a microscope slide and placing a cover-slip over it we can examine it under the microscope. It will be seen that this surface-layer consists of closely fitting cells which are colourless (Fig. 13, *B*, *e*); in broad leaves the outline of the cells, as seen in surface-view, is sinuous like the edge of a piece of a jig-saw puzzle (Fig. 14); in long and narrow leaves, such as those of a narcissus or a blade of grass, the cells are longer and relatively narrow and have straight walls. This closely fitting covering layer or skin, known as the epidermis, extends all over the surface: the outer walls of the cells directly exposed to the air are rather thicker than the inner walls, and over them is spread a thin transparent skin which serves much the same purpose as a layer of paraffin

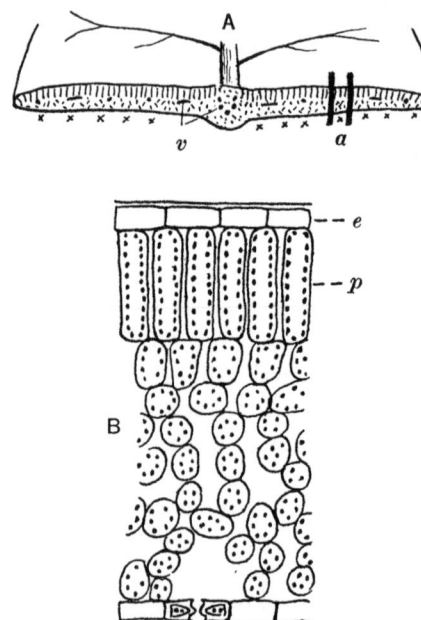

Fig. 13. *A.* A leaf cut across exposing the internal structure. *v*, veins in section. *B.* Part of the section (*A, a*) diagrammatically represented in *A* enlarged showing cells: *e*, of the upper surface-layer (epidermis); *p*, cells of the palisade layer with smaller cells below; *s*, one of the pores (stomata) on the lower epidermis. The crosses (×) on the lower margin of *A* indicate the position of the stomata. The black dots in the cells (*B*) are the chlorophyll-bodies.

or wax by reason of its comparative impermeability to water, water vapour, and air. Immediately below the upper epidermis is a layer of very regular cells which in the section of the leaf appear to be arranged like the boards of a palisade or fence (Fig. 13, *p*); each cell is much longer than broad, circular in section and so placed that its long axis is at right-angles to the leaf-surface. It is noteworthy that the cells of this palisade, as it is called, are not intimately connected one with another but most of their adjacent walls are separated by narrow spaces through which air can freely circulate. Each cell

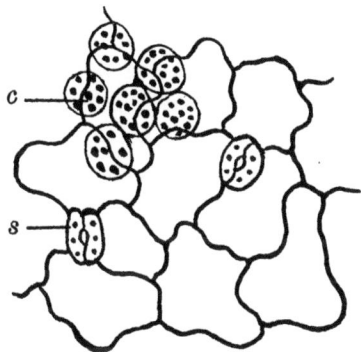

Fig. 14. Surface-view of a small piece of a leaf showing a few cells of the outermost layer (epidermis), two stomata (*s*), and the circular, upper ends of some of the palisade cells (*c*) seen through the transparent superficial layer. The black dots are chloroplasts.

contains living protoplasm lining the wall and the rest of the cell holds watery sap: in the protoplasm are numerous green bodies, known as the chlorophyll-bodies or chloroplasts, specialized, oval or spherical pieces of protoplasmic material holding, like little sponges, a green substance, chlorophyll.

Continuing our examination of the inside of a leaf we find that the palisade layer of cells rich in chlorophyll is in contact with a looser tissue composed of shorter cells and distinguished from the more compact layer above it not only by the smaller size of the cells but also by larger air-spaces which give the lighter colour to the underside of leaves. This is the structure characteristic of the great majority of leaves; a more compact and regular tissue next the upper face and a much looser and more spongy tissue next the lower face. Chlorophyll is present in the cells of the looser tissue but in smaller quantity than in the upper half of the leaf. Within the leaf there are also the veins composed of conducting tissue, in which relatively long pipe-like cells are the distinguishing feature; some of the tubes conduct water and others serve as channels for the transport of sugars and other soluble carbon compounds.

Returning to the epidermis we notice that in most leaves the superficial layer over the lower face is pierced by very small openings (Figs. 13, 14, *s*): these are known technically as stomata (Gk. *stoma*, a mouth). These openings play a dominant part in the operations of leaves and deserve a closer examination. Seen in surface-view each opening, or stoma, consists of two cells differing in form from those of the epidermis as a whole: the two cells are curved and may be compared with a couple of miniature sausages placed side by side, stuck together at the ends and then pulled apart in the middle so that an oval space is left (Fig. 14, *s*). In the stomata seen in surface-view in Fig. 14, *s* and in the stoma seen in section in Fig. 13, *B, s* the pores are open. This is only a very partial description of what is in reality rather an elaborate mechanism, though it will suffice for our present purpose. The two cells bounding the central pore are known as the guard-cells of the stoma. The sap of the guard-cells alters from time

THE GREEN LEAF

to time in the concentration of its contents and by so doing absorbs or loses water which causes the cells to expand or contract: expansion increases the curvature of the thin, elastic cell-walls and so enlarges the opening of the stoma, whereas in contraction the guard-cells come together and the stoma is closed.

Let us now endeavour to form an idea of the number of stomata and the size of the stomatal pores. As previously stated, on most leaves the stomata are much more abundant on the lower than on the upper face of a leaf and are often confined to the lower epidermis. On the leaf of a birch tree, for example, there are no stomata on the upper surface, while on the lower there are about 237 to each square millimetre. There are about 170 to each square millimetre on the upper surface of a sunflower leaf and 325 on the lower. The actual opening or pore is on the average not more than $\frac{1}{2500}$ inch in size. A hole made by a sharp needle is enormous in comparison with the size of a stomatal opening. Despite the very small size of the pore the combined openings occupy an appreciable part of the total leaf-surface: about three per cent. of the area of the lower surface of a leaf may be occupied by the pores. On a single sunflower leaf there are about 13,000,000 stomata.

"Through the pores of the open stomata the internal air-spaces of the leaf are placed in communication with the outer air. Thus if air is forced into the cut end of the stalk of a leaf dipped into water, bubbles may be seen rising from its surface. Or if the leaf blade is submerged and air is drawn *out* from the stalk by suction the blade is seen to become dark green as water, entering by the stomata, replaces the internal air which is being removed by the suction. If the stomata are completely closed neither of these effects will be shown. Thus in so far as they respond to different external conditions by opening or closing, the stomata may be said to control the passage of gases into and out of a leaf. This passage of gases in and out of the leaf is not, however, a mass flow of air, like the inbreathing and outbreathing of lunged animals, or like the flow of air under pressure in the experiment we have just described. It is just a result of the universal tendency of all gaseous molecules to spread or diffuse. If the concentration of carbon dioxide molecules is greater inside the leaf than in the outer air the net result will be a passage of molecules out of the leaf. Similarly when by the consumption of oxygen for respiration the concentration of oxygen molecules in the leaf is reduced below that in the outer air, oxygen molecules from the outer air will spread or diffuse into the leaf. Thus oxygen would be passing into the leaf through the stomatal pores and at the same time carbon-dioxide would be passing out. If the stomata are closed this gaseous exchange can take place only with extreme slowness, through the walls of the epidermis.

"Generally speaking the stomata are open during the day and shut, or partially shut, during the night. They tend to shut also when through excessive loss of water the leaves become limp or wilted. But to these as to most rules there are many exceptions."*

In a single day a birch tree gives off as vapour through its leaves about $15\frac{1}{2}$ gallons of water and on a very hot day as much as 80 gallons may be lost.

* The paragraphs within inverted commas are contributed by Dr Maskell.

THE GREEN LEAF

The giving off of water vapour from the wet cell-walls and its passage through stomata from the interior of the plant to the outer air is a process of evaporation as inevitable as the evaporation of water from any wet surface exposed to unsaturated air. In the plant, however, the rate of this evaporation is to some extent under control; the escaping vapour must pass through definite openings and the degree to which these are open or closed is under the control of the living protoplasm. The giving off of water vapour is spoken of as transpiration. Transpiration was formerly regarded as a process depending upon some vital activity; it is in fact simply a physical process, the price which must be paid by an organism exposed to the air.

Water passes from the ends of veins into the soft tissue of the leaf where it evaporates and eventually passes through the stomata; as evaporation occurs fresh supplies make good the loss. As the water evaporates from the wet cell-walls they become unsaturated and tend to suck up water from the cells between them and the veins of the leaf: these cells in turn draw water from the veins. In this way a tension is transmitted along the water-columns which stretch as continuous channels from the tips of the leaf-veins to the rootlets. We can best visualize transpiration as a pulling force by thinking of the ascending sap as a continuous column—a rod of water from one end of the plant to the other with its upper end in the leaves and its lower end in the roots. The column receives additions from the soil at one end, and in the leaves at the other end water is given off by transpiration: the strength of a rod of water is considerable and a force—such as the pull due to transpiration loss—acting upon it at the top is transmitted without breaking the continuity of the column. Further reference to this subject is made on p. 77.

It is not only aqueous vapour that passes through the stomata into the external air. The living cells of the plant are constantly giving off carbon dioxide in respiration; they respire as do all living cells whether animal or plant, and this product of combustion of the plant's fuel escapes through the stomata, or at least some of it does. On the other hand, it is through the stomata that the leaf receives supplies of oxygen from the outer air: the oxygen diffuses into the air-spaces between the leaf-cells and is thus available for setting into operation the essential chemical changes by which the energy stored in carbon compounds is liberated and rendered available as a factor in operating the whole plant machinery.

There remains to be considered the most important and most surprising service of the leaf in which the stomata take a leading part. It is through the stomata that the carbon dioxide of the outer air, which is to be used for the manufacture of sugars, passes into the leaf. As the carbon dioxide reaches the wet cells within a leaf it is dissolved in the water which saturates the cell-walls and lies on their surfaces, and thence passes in solution through the walls into the living protoplasm containing the chloroplasts.

CHAPTER VII

The Green Leaf (continued): How a Leaf utilizes Radiant Energy

We will now briefly enquire into the property of chlorophyll by virtue of which it enables the living plant to extract carbon from the air. Its chemical composition is by no means simple; it contains the elements carbon, hydrogen, nitrogen, oxygen, and magnesium. It is not a single compound which is green, but in addition to the green pigment known as pure chlorophyll or more correctly two green pigments, chlorophyll *a* and chlorophyll *b*, differing very slightly from one another in colour, it includes two other compounds, a yellow pigment and an orange-red pigment. It is the orange-red pigment which gives colour to the root of a carrot and contributes to that of the fruit of a tomato. In other words, the green colouring matter of leaves which is spoken of simply as chlorophyll is in reality a mixture of four pigments. A solution of chlorophyll can be made by placing a mass of green leaves in methylated spirit and warming on a water-bath.

The next step is to note the action of the chlorophyll solution on ordinary light, either light direct from the sun or from an artificial source, such as a bright flame. But before doing this let us briefly consider what we mean by light. The light from the sun travelling through space at the rate of 186,000 miles a second reaches the earth as radiant energy; much is absorbed by the atmosphere, which forms an envelope separating the earth's surface from the empty space beyond, and much is reflected back from the earth's surface. It has been shown that only about 30 per cent. of the total solar energy which reaches a leaf is actually absorbed: this is used in part by the chlorophyll and living protoplasm as the means by which the supply of carbon is kept up, and in part it provides the stimulus which directs and regulates growth. Light as we see it is white or colourless,

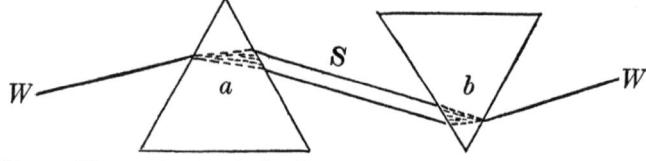

Fig. 15. Diagram showing the breaking up of ordinary, white light (*W*) as it passes through the prism *a*; also the re-conversion of the spectrum (*S*) into ordinary light (*W*) in its passage through the prism *b*.

but as Sir Isaac Newton proved nearly 260 years ago, when it is passed through a prism (Fig. 15) it is changed into a series of colours in the form of a rainbow band called a spectrum; at one end the colour is red and at the other end violet, with orange, yellow, green, blue, and indigo between. In its passage through the glass of the prism, light is broken up into its component parts: this is due to what is known as refraction. The beam of light (*W*) passes through the prism *a* and emerges as a coloured band or spectrum (*S*) including violet, indigo, blue, green, yellow, orange and red. The waves of light as they pass from the air into the glass, which differs from air in density, do not travel through the denser medium as one single beam, but

THE GREEN LEAF

the different components are separated or bent from the path of their course through the air in different degrees, and they emerge from the other side of the prism as separate waves which we recognize as the coloured bands of the spectrum. If the light, after the passage through a prism has spread its component waves into a coloured band, is then passed through a second, reversed prism (Fig. 15, *b*), it will be restored to its original colourless state (*W*).

Light is made up of a series of waves of different lengths and frequency, all travelling with the same velocity. If we tie one end of a rope to a support and move the free end up and down we impart a vibratory or wave-like motion to the rope; the frequency of the wave depends upon the rapidity of the movement given to the rope, and the frequency differences are recognizable in light as different colours and in sound waves as different notes. The expression wave-length means the distance between two points which occupy corresponding positions on the undulating length of a system of waves: the red rays are the longest and those at the violet end of the spectrum are the shortest, just as in sound the longer waves are low notes and the short waves high notes. A green leaf may absorb about 30 per cent. of the light which reaches it; not more than about 4 per cent. of this is absorbed by the chlorophyll. The light absorbed by the chlorophyll and used as a source of energy is mainly from the orange and red part of the spectrum. The violet and blue rays appear to influence the movements of plants in response to the stimulus of light. The most active rays in the process of carbon extraction are in the red portion of the spectrum.

In order to prove that chlorophyll absorbs certain rays of light we take a solution of it, pass light through it, and examine the spectrum of the light after it has traversed the green solution. It will be seen that this spectrum, which may be called the chlorophyll spectrum, differs from the ordinary or solar spectrum in the occurrence of certain black bands, cutting vertically across the rainbow colours, which are not present in a spectrum of light obtained by means of a prism. A black band or streak of darkness means the absence of certain parts of the light, the loss of certain waves. The most conspicuous black band is in the red part of the chlorophyll spectrum, and there are other bands which we need not consider further.

Experiments made with a chlorophyll solution or with the chlorophyll that is contained in the special bodies (chloroplasts) within the living protoplasm prove that this green substance absorbs some of the radiant energy from the sun. This then is the energy which, through the possession of chlorophyll, plants have at their disposal, a force enabling them to accomplish a feat beyond the power of animals, namely the pulling apart of carbon and oxygen in carbon dioxide gas.

It is difficult to believe that the minute chloroplasts, which are not more than 10μ ($\frac{1}{2500}$ of an inch) in diameter, are able to absorb an amount of light sufficient to enable a plant to manufacture the large bulk of carbon-containing compounds which we know it does. But if we reflect that there are about 400,000 chloroplasts on the average in a sample 1 sq. mm. in area cut out of a leaf, their co-operative efficiency is more readily appreciated. The total surface-area of the chloroplasts in a leaf with an area of

THE GREEN LEAF

10 inches (the leaf measured was taken from a castor-oil plant) is said to be at least 200 sq. yards. This apparently incredible statement needs a word of explanation. A large leaf contains millions of cells and in each cell are many chloroplasts. Can we then calculate the total area of the chloroplasts in a large tree? We can make an approximate estimate, and it must be remembered that this is hardly more than a guess based on evidence which does not provide all the data necessary for a scientifically accurate statement. Assuming that on a well-grown elm tree there are 7,000,000 leaves (a number taken from a published source): if we know the average area of an elm leaf we can calculate, on the basis of the estimate already given of the chloroplast surface-area in a leaf 10 inches square, the surface-area in an elm leaf and therefore in the tree as a whole. The following convenient method of measurement was adopted: it was carried out for me by one of the assistants in the Cambridge Botany School, Mr E. T. Scott. The outlines of five elm leaves were traced on a piece of paper: the tracings were cut out and each was carefully weighed: the average weight of a leaf was found to be 0·167 grm. The next step was to cut from the same kind of paper a piece 10 cm. square and weigh it: the weight was 0·783 grm. We know therefore that this weight represents an area of 10 sq. cm. and from the average weight of an elm leaf it is easy to calculate its area, namely 21·05 sq. cm. or 3·262 sq. inches. Multiplying this area by 7,000,000 we have 22,834,000 sq. inches as the total leaf area in the tree. In a leaf 10 inches square the chloroplast area is 200 square yards: from this we calculate that the total chloroplast area in an elm tree with 7,000,000 leaves is

456,680,000 yards, or 94,355 acres or 147·4 sq. miles! It is hardly necessary to point out that there must be a considerable range in the amount of light received by chloroplasts in different parts of the tree; but the estimate serves as an illustration of the important rôle of the almost infinitely small when we are dealing with enormous numbers.

Having overcome the very powerful forces which hold together the carbon and oxygen in carbon dioxide the leaf at once brings the carbon into combination with hydrogen and oxygen, and the first carbohydrate formed, which has a relatively simple construction, is rapidly converted into other more complex carbohydrates such as sugar and starch. Starch is a solid substance which is easily detected when the cells of a plant are examined under the microscope and a little tincture of iodine is added: the iodine gives a blue colour to starch and even the smallest grains are readily seen. The presence of starch in a leaf can be demonstrated on a large scale: take a leaf from a plant which has been exposed to the sun for a few hours or even a shorter time, bleach it by dissolving the chlorophyll in spirit and then place it in a saucer containing iodine. The leaf will very soon become dark blue or almost black owing to the abundance of starch in the cells. The experiment should be repeated with a variegated leaf in which white or yellow patches occur in the green ground: iodine added to such a leaf after exposure to light reveals the presence of starch only in the parts of the leaf which contain chlorophyll. A third experiment is worth performing: apply the iodine test to a leaf taken from the tree in the early morning before the sun is up. It will be found that little or no starch is present.

THE GREEN LEAF

Now what does all this mean? A leaf after exposure to light even for so short a time as an hour or less has produced starch; by absorbing light all day it is able to manufacture a large quantity of starch and other carbohydrates. One of the first carbohydrates to be formed is sugar, which being soluble is less easily seen than starch, a solid readily detected on the addition of iodine. Sugar is one of the chief sources of energy in the plant since it is by the constant breaking up of sugars into carbon dioxide and water that the living cells obtain the energy for their life-processes. Sugars and starch are convertible one into the other: we have seen that a leaf tested for starch in the early morning is practically starchless, whereas at the end of the period of daylight starch is plentiful. The starch, soon after its production, is changed into sugar, the solid starch becomes a sugary solution and in this form the carbohydrate is carried away from the green cells of the leaf through some of the pipes of the conducting veins to other parts of the plant where it is needed. The sugar in beetroot and in sugar cane, potatoes and grains of wheat is all made in the first instance through the agency of the light-absorbing chlorophyll and subsequently conveyed to the stems or other organs, where it occurs in sufficient quantity to be extracted for our use.

Fig. 16 shows part of a leaf highly magnified: light passes through the epidermis (one cell of which is drawn below the arrows), and is partially absorbed by the chloroplasts in the long palisade cell and in the smaller cells below. The soluble carbohydrates manufactured through the agency of the light-absorbing chloroplasts pass eventually into the sieve-tubes of the food-conducting

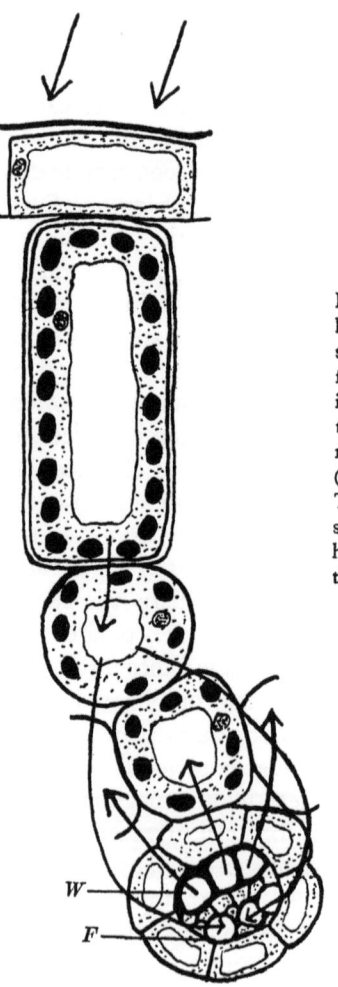

Fig. 16. Part of the inside of a leaf highly magnified. The two arrows show the direction in which sunlight falls on the leaf. The sugar produced in the palisade cell passes through the smaller cells and eventually reaches the food-conducting tubes (sieve-tubes: F) of one of the veins. The water-conducting tubes (W) supply the leaf with water which has travelled through the stem from the roots.

THE GREEN LEAF

tissue (phloem) seen at *F*. The water reaches the leaf from the roots through the woody vessels (*W*) of the conducting tissue which consists both of wood and phloem and forms the veins.

Before passing to a rather difficult subject suggested by the statement that starch and sugar are convertible one into the other, there is another aspect of the work done by the chloroplasts which must be noted. When the carbon dioxide combines with water to form sugars, oxygen is split off and escapes as a gas. In order to prove this we take pieces of a common water plant, as that is convenient for our purpose, and put them into a vessel containing water. A funnel is inserted into the vessel with its stem pointing upwards: the whole is exposed to light and after a short time bubbles will be seen to rise from the submerged leaves: these can easily be collected in a glass tube (test-tube) filled with water and placed over the end of the funnel stem (Fig. 17). Gradually some of the water in the test-tube will be replaced by the gas rising in bubbles from the plant. The test-tube is then removed by placing a thumb against the open end under water: on reversing the tube

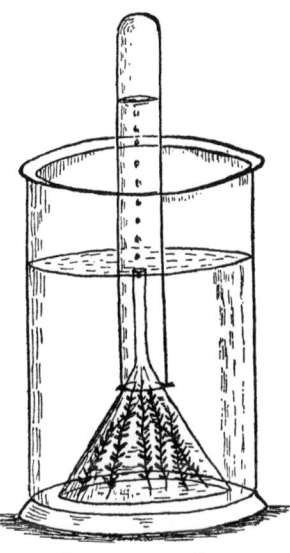

Fig. 17. Experiment to demonstrate that a green plant gives off oxygen in sunlight.

the water and gas are kept imprisoned in the tube by the thumb until, with the free hand, a glowing splinter of a match has been prepared by letting the match burn for a short time and then blowing out the flame. On introducing the glowing splinter into the gas we find that it bursts into flame and this proves that the gas is oxygen which was once part of the carbon dioxide in the air. Many years ago, in 1774, Priestley discovered that air which had been rendered foul by the breathing of animals, that is by the loss of its oxygen and the accession of carbon dioxide, could be made pure again by introducing into it a green plant. The precise explanation of this was not forthcoming until some years after Priestley's day.

The facts are these: green plants in light give off oxygen, but not in darkness. We have, however, stated on a previous page that all living plant cells like the cells of animals give off carbon dioxide. During the night it is very easy to demonstrate that plants do this, but in the daytime the liberation of carbon dioxide is masked by the reverse process, the giving off of oxygen. As the carbon dioxide is released in the combustion of the food in a plant's body the living chloroplasts at once seize upon it and retain the carbon: hence in daylight it is not so easy experimentally to prove that carbon dioxide is constantly being produced. It is natural therefore that misunderstanding should arise on the actual behaviour of plants: it is often said that plants give off oxygen, while animals give off carbon dioxide. In the daytime this is true, though not wholly true. Plants are always giving off carbon dioxide; that is to say they are always respiring like animals; they are always utilizing oxygen for burning up their fuel, but unlike animals they give off oxygen in sunlight. An example of this common confusion of ideas may be quoted: a pupil when asked what he knew about respiration and the extraction of carbon from carbon dioxide by green plants, replied that if a rabbit and a cabbage were placed under an inverted glass jar they might live for ever, the cabbage would keep up the supply of oxygen and the rabbit the supply of carbon dioxide!

In addition to the manufacture of carbohydrates, the sugars, starch, the cellulose of which cell-walls are made, and other similar compounds, the leaves of a plant also make proteins, those more complex bodies already mentioned: their production is a monopoly of living protoplasm; they have never been made by man. Proteins, such for example as the white of an egg, contain in addition to the constituents of carbohydrates, the element nitrogen and one or two other elements. It has not so far been possible to construct in detail the chain of events by which the plant converts carbohydrates and other carbon-containing compounds into proteins; but we know at least that the nitrogen is derived by the green plant from nitrogen-containing salts in the ground. The method by which the supply of nitrogen is kept up will be discussed later. Meanwhile, attention may be called at this stage to the conversion of sugar into starch, which is one of many operations characteristic of living organisms. Let us first give an example: starch makes its appearance as small grains in the chloroplasts of a leaf some minutes, it may be, after exposure to light. The starch grains continue to increase during the day, but most of the starch disappears during the night. What happens to the starch? It is converted into the soluble sugar and carried away. We have implied by what has been said that sugars and starch are made only in green leaves: this is true in a sense. It is only in the green parts of plants, leaves or green stems, that carbon can be obtained for carbohydrate production: we know, however, that if we require starch in quantity we take a potato or a grain of wheat or some other grain. What happens is this: starch is formed, probably from sugar, in the leaf cells; it is then changed back into sugar and as such is carried either to the developing potato tuber underground or to the ripening ear of corn and then changed back again into solid starch which, as it accumulates, forms comparatively large grains. In the cells of that part of a potato which we eat starch occurs as large grains, made up of concentric layers which have been gradually deposited round the particle first formed within a small body similar to the chloroplasts of a leaf through lacking chlorophyll.

Without attempting to describe the nature of this process, it is worth while to say a little about it, because the conversion of starch into sugar is one of many instances of a special kind of chemical change which is one of the distinguishing features and a most important characteristic of living organisms. In every seed there is a store of food left by the parent plant as capital on which the seedling draws until it is able to make its own food. Food to be available in the different parts of a plant's body must be in a state in which it can be carried from one cell to another. The solid starch in a seed is useless until it is changed into sugar which is easily transported. The tip of a young seedling has a sweet taste because it is rich in sugar which has come from the starch. The living cells produce a substance which is technically known as an enzyme (Gk. *en*, in; *zume*, leaven) and was formerly called a ferment (Lat. *fervere*, to boil). By the action of a particular enzyme starch is converted at ordinary temperatures into sugar; the molecules are rearranged and water is added to the starch. This enzyme, which is one of many, is called diastase (Gk. *diastasis*, separation) and can be extracted from plant cells and its action demonstrated in the laboratory. Other familiar enzymes, such as those in our saliva and in various parts of the body, take a leading share in what we call digestion: they cause food to disintegrate into diffusible substances which the blood can absorb. It is well known that yeast produces fermentation and is able to convert sugar into alcohol. This was formerly thought to be a property inherent in the living yeast plant, which was therefore called a ferment. We know now that it is possible by pounding up yeast cells in a mortar to extract something from them which is soluble in water and, even after dissociation from the living protoplasm of the plant, can produce alcohol from sugar: this is an enzyme. When the yeast enzyme causes the conversion of sugar into alcohol, one of the results of the change is the evolution of carbon dioxide: it is this production of a gas which has led to the employment of yeast in making bread. The gas liberated during the action of the yeast makes the bread rise. In cheese-making use is made of another enzyme, known as rennet, obtained from the stomachs of calves and pigs: this by acting on milk forms the curd.

Enzymes have many remarkable properties: they are

not themselves alive but are produced by living cells; they are able to accelerate chemical changes enabling them to occur at appreciable rates at ordinary temperatures whereas in the laboratory the same rapid rate of change would require a very high temperature.

An illustration* may make this clearer: a plate of glass which has been carefully cleaned is placed in a sloping position, one end on a table, the other resting on a support which can be raised or lowered. A weight, the base of which has been polished, is placed on the upper end of the sloping plate; at a certain angle of inclination the weight slides slowly down. It moves slowly because the friction of the weight on the glass retards the action of gravity. If some oil is smeared over the bottom of the weight, it slides down rapidly. The oil accelerates the descent of the weight by reducing friction; the oil in its accelerating influence is comparable with an enzyme. Enough has been said to call attention to the enormous advantage given to organisms by enzymes: they are a necessary accompaniment of life and through their influence solid substances rich in potential energy are rendered available as sources of nutrition.

* Taken from *Principles of General Physiology*, by W. M. Bayliss.

CHAPTER VIII

Roots and what they do

Though repeated reference has been made to roots and their absorption of water from the soil we have not paid attention to their structure or methods of work. Roots serve a double purpose: they anchor a plant in the ground —a mechanical service—and they actively take in water, though it is only the younger and more delicate parts of a root-system which do this. The actual tip of a root is covered by a protective cap of cells (the root-cap) which is constantly worn away on the outside by rubbing against solid particles in the soil. The loss is made good by the formation of new cells on the inner surface of the tissue which is constantly being destroyed on the outside: these repairs are effected by the permanently young tissue at the tip of the root which is covered by the cap. A short distance behind the tip the surface of the root is covered with a felt of delicate hairs as can readily be seen if we sow seeds of mustard or cress on damp blotting paper or flannel, and then examine the seedling roots. The delicate roots are seen to be covered over for part of their length by numerous fine threads, each of which is a hair formed by the outward extension of a surface cell. It is these living tubular hairs, the root-hairs, which absorb water; as the root grows longer and wanders further in the ground the old hairs shrivel up and new ones are regularly produced, always on the region of the root near the apex. Thus, as new ground is invaded a new set of hairs is ready to tap the fresh stores of water.

If we pull up a young plant and shake the roots it will be found impossible to get rid of all particles of soil because some of them are firmly attached to the substance of the walls limiting each hair.

Soil is made up of particles of sand, clay, and many other substances mixed with the dark material known as humus, which is formed by the disintegration and decay of organic matter. In wet soil the spaces between the solid particles are more or less filled with water and in very wet soil there is often a deficiency of oxygen through the displacement of the air by water. The roots of a plant must have access to oxygen or their vitality is destroyed. In addition to the free water in the interstices there is a film of water over each solid grain which disappears only in very dry weather. As the young branches of a root push their way through the soil the water adhering to the solid particles, with some of the free water, passes through the permeable walls of the hairs into the inner tissue and eventually reaches the conducting pipes of the woody strands in the central part of the root. The water holds in solution various salts though only in small quantities: within each root-hair is a lining of protoplasm next the cellulose wall, the rest of the hair being occupied by watery cell-sap. Between the sap and the water in the soil there are therefore two membranes, the cell-wall and the layer of protoplasm. The water passes through the two membranes because the cell-sap attracts it; it exerts a force which causes the water to pass through both the cellulose wall and the living protoplasm. We can illustrate, partially at least, the sort of thing that happens by taking a bladder and filling it with water containing sugar: into the bladder is inserted a long glass tube which is tightly tied into position. The bladder is now immersed in a vessel containing ordinary water. After a time the sugary water will be seen to rise in the tube and this shows that the volume in the bladder has been increased by the inflow of water from without. This is the phenomenon known as osmosis (Gk. *osmos*, thrust, push). If the bladder were a closed bag it would increase in size and become rigid like an inflated football. It is in a similar way that the root-hair cells tend to keep fully extended when in contact with the soil water. From the inner surfaces of the root-hair cells water is continually passing into the conducting strands owing to the tension transmitted from the leaves: this leads to a corresponding intake of water by the root-hairs from the soil. Water passes readily through the cell-walls of the root-hairs and through the protoplasm, but the latter forms an efficient barrier to the passage of certain other substances. Sugar, for example, occurs in the cell-sap, but it is not normally drawn out of the plant even if a strong salt solution is placed outside the root-hair: the stronger solution withdraws water from the cells by osmosis, but the sugar is retained within the protoplasm so long as this remains uninjured.

A good example of the control exercised by living protoplasm on the passage of substances through it is afforded by the cells in a red beetroot. If we cut a very thin slice from a beetroot and place it in a little water on a microscope slide and then put a cover-slip over it, it is readily seen that the red colour is due to some colouring matter dissolved in the cell-sap. On replacing the water in which the thin slice of tissue is mounted by a solution of salt, or some other substance, which is stronger than the cell-sap, we can follow the gradual withdrawal of water from the cells and the consequent collapse of the protoplasm which leaves the cell-wall and forms a ball. On replacing the salt solution by ordinary water it is seen that the cells regain their normal size and the protoplasm is pushed back against the walls as the volume of the sap increases. The important point is that though much water was withdrawn the red colouring matter did not escape: it was not allowed to pass the protoplasmic barrier. When the red cells are killed the coloured sap at once escapes: this shows that dead protoplasm is very different from protoplasm that is alive.

In addition to the intake of water by the roots as a result of transpiration, there is another phenomenon involving uptake and upward movement of water through the stem which is known as root pressure. If a stem of a plant be cut across a short distance above the ground, particularly when the sap is rising vigorously in the spring, liquid will be seen to ooze from the surface: the force with which it exudes affords a measure of the energetic action of the hairs borne on the several branches of the root-system. This bleeding of stems, as it is called, is evidence of a pressure exerted by the roots. This force-pump action has often been regarded as one of the primary causes of the flow of sap through the wood of a tree, but it is not shown by all plants to any appreciable extent nor at all seasons. It is by no means the chief factor in keeping up a regular supply of water to the tree as a whole.

ROOTS AND WHAT THEY DO

We must now supplement a little the reference previously made to the rise of water in stems: it is a subject which has long puzzled botanists, and for the following reason among others. It is a common experience to find that many of the most familiar phenomena in nature are not thoroughly understood. As John Donne wrote many years ago:

> Why grass is green, or why our blood is red,
> Are mysteries which none have reach'd unto.

We know that water travels through the cavities of the small wooden pipes which serve as conduits and it must be remembered that these are dead and therefore the protoplasm cannot, one would suppose, play any part in lifting the water. Some years ago it was believed that certain living cells, which are associated here and there with the dead conducting tubes, acted as pumping stations; but it was proved experimentally that a tree in which the living tissue had been killed by a poisonous solution supplied to the roots was still able to conduct the sap to the topmost branches. The probability is that for the most part at least the chief factor concerned is a physical and not a vital force.

" It was at one time thought that the maximum height to which water could rise, by purely physical processes in the trunk of a tree, would be the height at which water would stand in a water barometer. Since water is about one-thirteenth as dense as mercury this height will be about 13×30 inches; the value is actually 33·7 feet for normal barometric pressure. This height is obviously far short of what is required in tall trees which may reach 300 feet. It may, however, be urged that we are not justified in assuming that the height reached in the narrow conducting tubes of the tree will be the same as that reached in the very much wider barometer tube. It is well known that if an open tube is dipped into water the water will rise in the tube, above the level of the water outside, to a height depending on the diameter of the tube. We must, therefore, add the height of this capillary rise on to the normal barometer height to find the height to which water will rise in the conducting tubes of the tree trunk if these are acting like narrow-bored barometer tubes. The conducting tubes in a deciduous tree, e.g., a lime, have a diameter of $\frac{1}{500}$ inch; the smaller tubes in a pine tree are about $\frac{1}{2500}$ inch in diameter. The capillary rise for these narrower tubes would be 10·3 feet so that our corrected barometer and capillary height becomes, at most, 44 feet. Thus the capillary effect due to the narrow bore of the conducting tubes contributes very little towards explaining the tremendous discrepancy between the actual rise and that calculated as due to purely physical forces. For some time it was thought, therefore, that some pumping action on the part of the living cells of the tree must be involved in the ascent of water. In regarding the conducting tube as a water barometer we have, however, made one very important mistake. In the ordinary barometer there is an apparently empty space (Torricelian vacuum) above the mercury level. This space contains only mercury vapour and the column of mercury is maintained by the difference in pressure between the outer air, acting on the lower end of the barometer tube, and the negligibly small pressure of mercury vapour in this space. Now we have tacitly assumed that in the conducting tubes of a tree a similar vapour space will be formed whenever the height of the tube is greater than 44 feet. But in order to pull asunder a column of water, and so form a water-vapour space, a force not of one atmosphere but of many hundreds of atmospheres would be required, so that the formation of a vapour space in the conducting tubes must be a relatively rare occurrence. We must therefore visualize the ascending water as a series of *continuous* columns stretching from the soil water, through the tissues of the root, along the conducting tubes of the trunk and so on through the veins and through the fine pores of the walls of the leaf-cells right up to the surfaces where these cells are in contact with the internal air-spaces of the leaf. It may be asked why the weight of these long columns of water does not pull the water out of the cell-walls and so admit air into the upper ends of the conducting tubes. The pores in the cell-walls are, however, so fine that their capillary force is extremely great. They will retain the water even against a pull of some hundreds of atmospheres. Thus it is, after all, capillary forces that maintain the long columns of water in tall trees, but these capillary forces are those of the cell-walls with their fine pores and not those of the conducting tubes themselves.

We must now see how the water in these long columns is set in movement. If there were no loss of water vapour from the plant and no growth there would be no upward movement. Evaporation of water from the wet cell-walls into the air spaces of the leaf is, however, nearly always proceeding since the outer air is rarely saturated with water vapour. As the cell-walls lose water by evaporation the liquid water tends to retreat slightly into their fine pores. This creates a pull or tension which is transmitted along the whole length of the long columns of water right down to the soil water. So along the whole length of the long path water is set in upward movement and usually a state of affairs is reached in which upward movement just balances water loss by transpiration."*

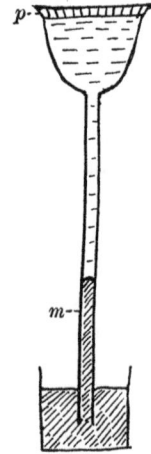

Fig. 18. An experiment to illustrate the upward pull exerted by an evaporating surface: *p*, plaster of Paris in the mouth of the funnel; *m*, column of mercury.

A simple illustration may help to make clearer the mechanics of ascending sap. A thistle-funnel—a long and narrow glass tube with a wide funnel at one end—is closed at the broad end with a plate of plaster of Paris (Fig. 18, *p*): when that has set the tube is filled with water, which has been boiled to drive off the air, and the narrow end inserted into mercury. The water evaporates through the plaster roof and a column of mercury (*m*) rises to take its place, being pulled upwards as the water is pulled upwards in the tree.

The rate of ascent of water in a branch with fresh leaves may be measured by the simple apparatus shown in Fig. 19. The cut end of a branch is forced into a short piece of stout rubber tubing (*C*) and tied firmly into it; the other end of the rubber is tied to the end of the short arm of a glass tube. Into the lower end of the two-armed tube is

* The paragraphs within inverted commas are contributed by Dr Maskell.

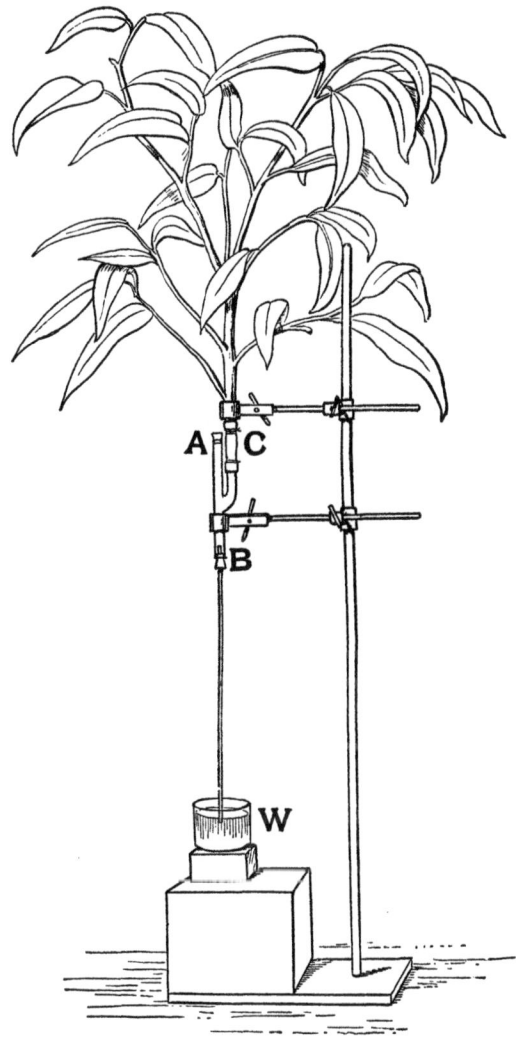

Fig. 19. An apparatus for demonstrating the effect of leaves on the ascent of sap in stems. (After F. Darwin.) For explanation see text.

inserted a cork which has been bored to admit a much narrower tube with a smaller bore (*B*). The tubes are then filled with water and the lower end of the capillary tube is dipped into a vessel of water (*W*) placed on a wooden block. When the leaves are vigorously transpiring, take away the wooden block and wipe the lower end of the capillary tube with a piece of blotting paper and quickly replace it in the vessel of water. It will then be seen that a small bubble of air caused by wiping the end of the tube is travelling upwards, and its rate can readily be measured by timing its ascent between two marks made on the tube. The rate of ascent of the bubble indicates the rate of flow of the water: the amount of water which passes upwards is approximately equal to that lost by evaporation from the leaves.

CHAPTER IX

Our Debt to Bacteria: the World's Supply of Nitrogen

We have enquired into some of the activities of green plants by which raw material is worked up in the living laboratories into more and more complex substances which provide the energy for running the machine, and have seen that it is only green plants which are able to convert simple raw material into carbohydrates and proteins by reason of their unique power of utilizing radiant energy. The following anecdote related by a Russian botanist illustrates the significance of this property of green plants. George Stephenson in the early days of the locomotive asked his friend Buckland, as they watched the progress of a train, if he could tell him by what force the engine was driven. Buckland replied that the force was supplied by one of Stephenson's engines driven by an engineer. "No," said Stephenson, "it is sunlight. It is light that has lain stored in the earth for many thousands of years; the light absorbed by the plant during its growth is essential for the condensation of carbon, and this light, which has been buried in the Coal Measures for so many years is now unearthed and, being freed again as in this locomotive, serves great human ends." There remains to be considered the source and supply of nitrogen, the element which enters into the composition of all proteins and is therefore essential to all animals and plants. It has been said that the vast supply of nitrogen in the air is not utilized by green plants; though this is generally true there are a few partial exceptions of great interest. The word *partial* is deliberately used because the few green plants which are able to obtain nitrogen from the air do not do so unaided; they have in partnership with them certain minute organisms, Bacteria, which render inestimable service as intermediaries in the complicated process of gaining nitrogen from air. Before describing the nature of this partnership and the part played by the bacteria we will first consider green plants as a whole in their relation to nitrogen.

The great majority of green plants obtain nitrogen from salts of nitric acid, that is from nitrates, or from compounds of ammonia present in the soil in a form available to the roots. The first question is, where do the nitrogen-containing substances come from? The original source of nitrogen is the atmosphere, but the gas in a free state is not used by plants; how then does it come to be available in a combined state? Electric discharges during thunderstorms cause nitrogen to combine with oxygen forming oxides of nitrogen, and these with water produce nitric and nitrous acid, respectively HNO_3, HNO_2. This is one source, though a very insignificant one, of nitrates, which are salts formed from nitric acid. In Chili there are large deposits of nitrate of soda, the excrement of millions of sea-birds. The most abundant nitrogenous material on the earth's surface is in the form of complex organic compounds derived from the remains of animals and plants, the refuse of the living world. This is not to any appreciable extent available to roots as a source of nitrogen. The simpler nitrogenous substances such as salts of nitrogen, with the exception of the small quantity produced by electrical discharges in the air, do not occur in nature as products of chemical change unconnected with living organisms. Most mineral substances in soil are formed by the ordinary chemical action responsible for the wearing away and weathering of rocks, but the nitrogen salts have a different origin and their production is due to the activities of various kinds of bacteria; they are therefore organic and not inorganic products.

Soil consists not only of grains of sand, pieces of clay and other mineral matter, it contains also the dead remains of animals and plants, decaying leaves and twigs and decomposing bodies of innumerable animals that are rich in carbohydrates and nitrogenous compounds. The soil is not a mere mixture of dead materials; it is a world thickly populated by organisms. The earthworm is one of the larger inhabitants and a very useful member of the soil community: by passing soil through its body it aerates the surface-layers and helps by its digestive juices to decompose organic matter. By far the most important dwellers in the soil are bacteria and other microscopic organisms, some animal and some plant. It is estimated that in the upper layers of the soil there are from two to fifty millions of bacteria in each cubic centimetre.

Many people think of bacteria as pestilential, disease-producing invisible microbes which do more harm than good and are neither animal nor plant. It is true that in some respects bacteria differ from all other organisms; they have, however, as great a claim to be called plants as animals: bacteria on the whole are beneficial rather than harmful and, as we shall endeavour to show, some of them supply green plants with an essential article of food, the nitrogen, on which the whole living kingdom depends. It is often said that bacteria were among the first, possibly the first, inhabitants of a world that was previously lifeless: we know that bacteria have existed since the earliest stages in the evolution of the plant-world. The largest kind of *Bacterium* or *Bacillus* is visible only under a high magnifying power; there are still smaller organisms, presumably related to bacteria which are invisible under the highest power of a microscope, though there is conclusive evidence of their existence. One may say that the average *Bacterium* is about $\frac{1}{25000}$ inch in size. In shape, bacteria may be compared with a billiard ball, a cigarette and a corkscrew: most of them possess extremely delicate threads which by active vibration enable the organism to move. Various names are given to bacteria; the rod-like forms are known as *Bacterium* and *Bacillus*; others, spherical in shape, are called *Micrococci*, and so on. None possess chlorophyll. They were first discovered by a Dutch naturalist in 1683, who found them on magnifying material taken from human teeth and called them animalcules. Among the more striking characteristics of bacteria are their extraordinary power of endurance under conditions that are fatal to other organisms, and the rapidity with which they multiply. It is known that in some bacteria, e.g. the species which is the cause of cholera, the parent cell, by division into two, may produce a pair of new individuals once every twenty minutes: at this rate and assuming no mortality one hundred tons of solid bacterial

THE WORLD'S SUPPLY OF NITROGEN

matter would be produced in a single day. We shall confine our attention to bacteria which minister by their activities to green plants and therefore to the whole animal creation which lives directly or indirectly on plants.

We will first give a short account of the tremendously important service rendered by some of the bacteria in soil which manufacture simple compounds of nitrogen from the more complex carbon-containing nitrogen substances which the green plant does not use, or only in rare instances and with difficulty. The litter of plant *débris* and animal refuse in the soil undergoes decay: by this we mean that the vegetable tissues, consisting of cellulose with other relatively complex substances, and the nitrogenous compounds abundant in organic matter are gradually pulled to pieces by hosts of bacteria, which derive the energy which is as essential to them as to all other organisms by their power of decomposing complex carbon-containing and nitrogenous bodies into simpler inorganic substances. In this process of decay one of the products is the gas, carbon dioxide: some of this is dissolved in soil water and increases the solvent power of the rain as it sinks into the ground and comes into contact with minerals. The greater part escapes into the atmosphere and maintains the carbon dioxide content of the air. Another product of bacterial action is ammonia, the pungent gas which gives a characteristic odour to stables where the nitrogenous substance urea undergoes disintegration into simpler substances. Ammonia is similarly produced in the soil where animal refuse and other nitrogenous materials are broken down. It consists of one atom of nitrogen combined with three atoms of hydrogen (NH_3). Though ammonia may in certain circumstances be directly absorbed by roots the main uptake of nitrogen is in the form of nitrates.

The next stage therefore is the conversion of ammonia into nitrous acid which has the formula HNO_2; it contains only two atoms of oxygen whereas nitric acid (HNO_3) contains three. The formation of nitrous acid is the service rendered by one kind of bacterium known as *Nitrosomonas*, an exceedingly minute spherical cell. Without including in this general account a description of all the stages between the formation of ammonia and the nitrogen salts, special attention is drawn to the fact that the production of nitrates, which are salts of nitric acid, is taken over by a second specialist among bacteria, a form known as *Nitrobacter*. This converts the nitrous acid or its derivatives into nitric acid which in the form of its salts (nitrates) is absorbed by the roots of green plants. The outstanding facts are: the co-operation of different bacteria in the chain of events represented by the conversion of organic nitrogenous substances into soluble nitrates, in which the nitrogen can be secured through the absorptive action of root-hairs.

Many different kinds of bacteria have a share in the chemical changes which are the cause of the disintegration or breaking up of material that we speak of as decay and putrefaction: in the course of these changes ammonia and carbon dioxide are produced as two of the simpler substances characteristic of bacterial activity. But the next acts in the complicated process, by which nitrogen is combined with other elements in a form acceptable to green plants, cannot be accomplished by the bacteria

THE WORLD'S SUPPLY OF NITROGEN 89

concerned in decay: two other bacteria, *Nitrosomonas* and *Nitrobacter*, take over the work. The complete series of events from the first attack on the accumulated organic refuse to the final production of nitrates affords an admirable example of team-work among the smallest members of the plant-kingdom.

We pass now to another illustration of the way in which bacteria minister to the needs of the higher plants. The history of this second example of bacterial co-operation goes back to the days of Pliny and Virgil. Roman farmers found by experience that if a crop of beans, vetches, or certain other related plants had been grown in a field before planting wheat a better yield of the cereal was obtained. For many centuries before the scientific explanation was discovered it was known that crops of beans, peas, clover, vetches and other food-plants belonging to the pea family (Leguminosae) enriched the ground. We now know that they enrich it by adding to the store of nitrogen, whereas other crops, such as cereals, reduce the amount of nitrogen in the ground. The cause of this remarkable property of what are called leguminous plants, that is members of the pea family, was first demonstrated about forty years ago. If we examine the roots of a bean, a lupin, clover or other leguminous plant, we see here and there small swellings or nodules (Fig. 20) which were long ago described as galls, as in a sense they are. These nodules are

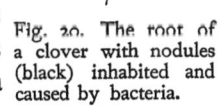

Fig. 20. The root of a clover with nodules (black) inhabited and caused by bacteria.

developed by roots in response to a stimulus given by invading bacteria. There are certain minute rod-like bacteria in the soil which are able to pierce the walls of the root-hairs of leguminous plants; they then make their way to the inner part of a root and their presence stimulates the plant to produce more tissue, and swellings are formed. The bacteria live comfortably and multiply vigorously in the cells of the nodules, utilizing as a source of the energy necessary for their life the carbon-containing compounds such as sugars which the green plant provides. By means of the energy thus obtained the bacteria are able to extract nitrogen from the air that is always present within the roots, and by bringing it into combination with other elements they eventually produce complex nitrogenous compounds. Thus, the beans, peas and other plants with which these nitrogen-fixing bacteria live in mutual, beneficial partnership obtain a store of nitrogenous material, while at the same time the bacteria derive nourishment and energy from the carbohydrates provided by the green plant. When the bean or clover crop has been harvested much of the nitrogen taken from the air by the bacteria is left in the ground with the dead roots and thus enriches it. It is interesting to note that if peas or other leguminous seeds germinate in sand which has been exposed to a temperature high enough to kill the bacteria and is therefore sterilized, the roots do not form nodules: this formation of nodules is evidence of the presence of certain bacteria which possess a most valuable quality, lacking in higher plants, namely the ability to make use of the free nitrogen of the air.

THE WORLD'S SUPPLY OF NITROGEN

Were it not for the services which bacteria render, the surface of the earth would soon be covered with an

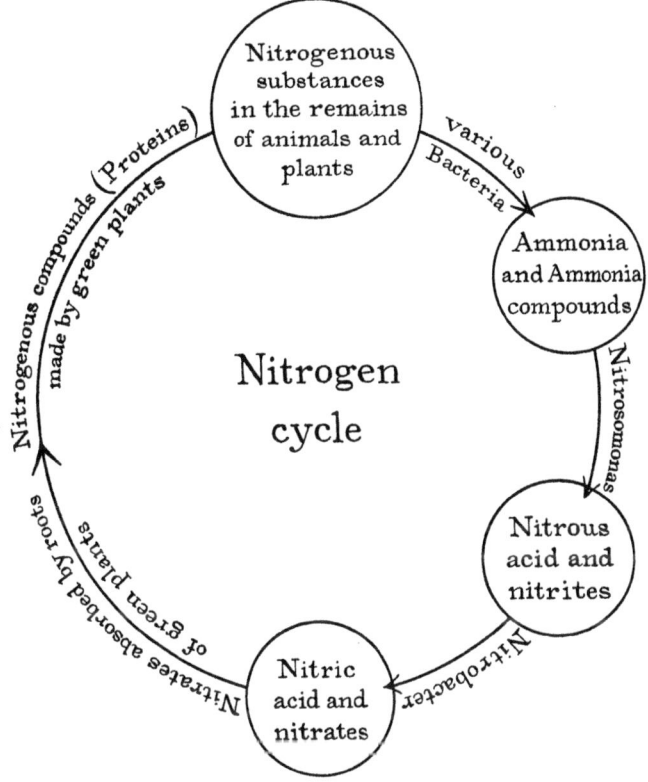

Fig. 21. The Nitrogen-cycle.

enormous mass of dead animal and vegetable matter: there would be an accumulation of millions of tons of nitrogenous organic material, but the nitrogen would be useless to green plants because they must have it in the form of simple salts. The diagram (Fig. 21) on the previous

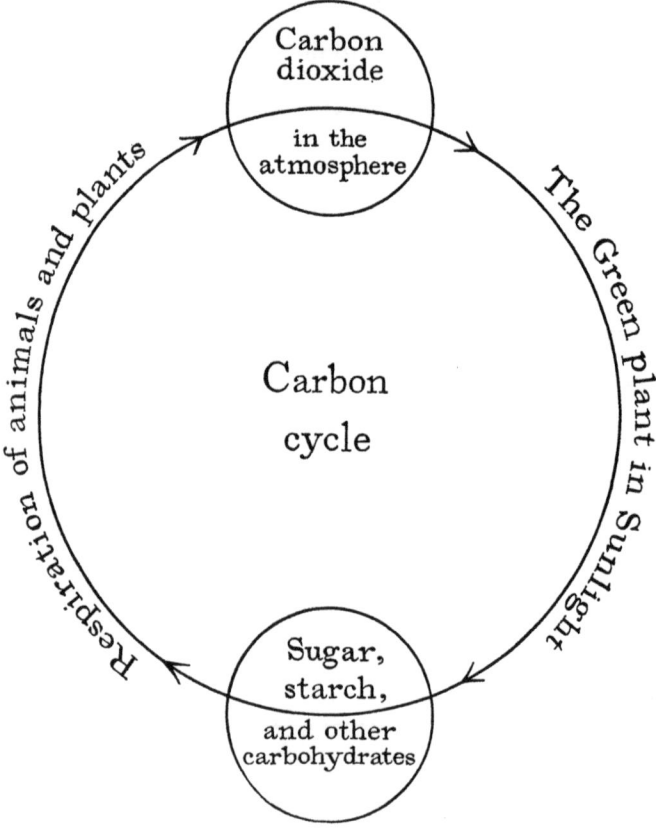

Fig. 22. The Carbon-cycle.

page illustrates the Nitrogen-cycle and a diagram of the Carbon-cycle is added for comparison (Fig. 22).

THE WORLD'S SUPPLY OF NITROGEN

The two diagrams (Figs. 21 and 22) represent in a simplified form the circulation of nitrogen and carbon in the world; they are intended to express the important facts, stated in this and the preceding chapters, (i) that green plants are the agents by which the carbon compounds (carbohydrates) essential to animal life are manufactured, and (ii) that it is through the agency of various bacteria that nitrogen is made available for the nutrition of plants.

CHAPTER X

Seeds and Seedlings

Having obtained some insight into the general plan of construction of a green plant, the methods by which the living cells obtain the raw materials and work them up into the complex substances which we use as food, and the co-operation of bacteria in maintaining the world's supply of usable nitrogen, we are in a better position to appreciate the initial stages of development of a plant as it emerges from the seed.

One of the most striking characteristics of the higher plants, which distinguishes them from the humbler members of the plant-kingdom, is the possession of seeds. By the higher plants we mean both those which produce flowers (Flowering plants), using the term flower in the ordinary everyday sense, and plants which from the production of cones are called Conifers, e.g. pine, fir, larch, etc. An oak and a pine tree, to take one example from each of the two great classes, reproduce their kind by means of seeds which become detached from the parent, and after a longer or shorter period of rest and, it may be, after being carried by wind or by animal agency some distance from the plants which bore them, germinate and produce seedling trees, shrubs or herbs. Ferns, mosses and the still simpler plants reproduce by spores, tiny cells which are scattered like dust by the wind. A seed and a spore are very different things: a spore is a very small,

SEEDS AND SEEDLINGS

single cell which becomes detached from the parent; it is often endowed with remarkable powers of resistance to unfavourable conditions, and on germination it gives rise directly to a new organism, or to some structure which represents a stage in the life-history of the parent plant. The protoplasm of a spore like that of a seed contains within it some directive forces, which on germination lead inevitably to the production of a definite type of plant-body: a spore does not contain an embryo, but it contains something which causes it to develop along pre-determined lines.

A seed is a relatively large and bulky structure made up of thousands of cells and within it is an embryo, a miniature plant-body in which the future root and shoot are already clearly defined. It will be easier to obtain a correct idea of the nature of a seed if we very briefly consider its development, though without going into detail. Two common flowers may conveniently be used in illustration of the origin of a seed, a buttercup and a primrose; though dozens of other flowers would be just as good for our purpose.

In the centre of a buttercup several green bodies are arranged in a spiral over the domed surface of the apex of the flower-axis; these are the carpels, that is, small cases or ovaries each containing a much smaller body known as an ovule. The carpels together are spoken of as the pistil or the female part of a flower. In a primrose (Fig. 23) the main part of the female organ consists of a single and relatively large green spherical case, or ovary, which lies at the bottom of the narrow, yellow tube of the flower: if this is opened a mass of small white ovules will be seen. An ovule when taken out of a carpel on the point of a needle looks like a small egg of an insect and it was for this reason that the name ovule was first used. The ovule of a flower when it has reached maturity contains within it an extremely small cell, the egg or female reproductive cell, which is ready to receive the male reproductive cell. It is unnecessary to describe the structure of an ovule; the essential thing to remember is that it

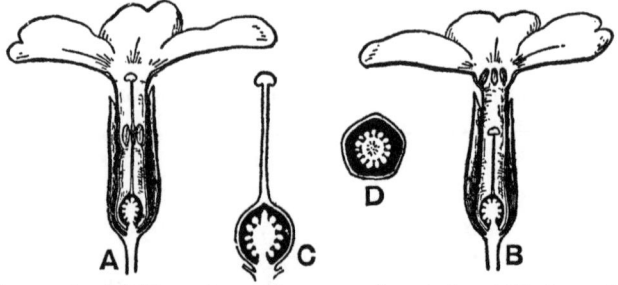

Fig. 23. *A* and *B*. Two primrose flowers cut through the middle: in one (*A*) the pollen-bearing organs (stamens) are attached to the yellow tube half-way along its length; in *B* they are at the top of the tube. In *A* the knob (stigma) at the end of the slender column (style) which rises from the spherical ovary is at the upper entrance to the tube; in *B* it is shorter. *C* shows the ovary (seed-vessel) cut through the centre with the small ovules (immature seeds) attached to a solid boss; *D* is a cross-section of the ovary. (After Yapp.)

contains, within a comparatively large and elaborate mass of cells, a single egg, the female cell. We are not now concerned with the method by which the male and female cells are brought together in the act of fertilization. After fertilization, and as the result of stimuli consequent on the fusion of the male and female cells, the ovule grows larger and its various tissues undergo considerable change and expansion: the ovule becomes a seed. The egg grows rapidly after fertilization and eventually forms the embryo: when a certain stage in development is reached growth stops, and the seed is said to be ripe. As the embryo is being formed the outer region of the ovule becomes harder and develops into a seed-coat.

A seed does not consist only of an embryo and a protective coat, which is often a hard shell: an essential feature is the presence somewhere in the seed of a store of food. During the development of an ovule into a seed building material is required for the numerous operations in progress, and this is supplied by the parent-plant; but the supply is more than sufficient to meet the immediate demands for constructive purposes. There is a substantial surplus which is reserved for use in the future when the embryo develops into the seedling. This reserve food reaches the growing seed as a supply of fluid material carried to it through the strands of conducting tissue which form a continuous system from the green leaves, where the food is made, to the flowers and all other parts of the plant; it is usually, however, not stored as a liquid but in a solid form. The conversion of the food from a soluble state, such as sugar, into a solid such as starch is effected by enzymes as described in Chapter VII.

It is worth noticing that while seeds always contain a store of food this is not invariably in the same place: in many seeds it is outside the embryo; in others it is stored within the embryo. An examination of two seeds will make this difference clear. Take a pea from a ripe pod and remove the outer envelope: within it is the embryo which consists almost entirely of two thick bodies, each flat on one side and rounded on the other: these are the two substantial seed-leaves, or cotyledons as they are called, of the pea embryo: on closer inspection it will be seen that they are attached to a very small axis representing, in miniature, the main root and the main stem of the future plant. It is easy to see that a broad bean seed is similar in structure to a pea. After soaking a bean in warm water we can readily peel off the coat and lay bare the large cotyledons of the embryo. In both the pea and bean the embryo contains a store of reserve food as starch and protein, in the cells of the seed-leaves.

A second type of seed is illustrated by the castor oil plant, the date palm, wheat and other cereals. We will examine the seed of a date which is the "stone" of the fruit. The edible flesh of the date is the enlarged wall of the ovary, or carpel of the flower. Along one side of the stone is a groove; the opposite side is rounded (Fig. 24, I, II). If we look carefully along the rounded side we see near the middle a small circular spot about the size of a pin's head; by sticking the point of a pin into the spot we notice that it is soft and very different from the rest of the stone from which it can be separated as a distinct little body, the future date palm. Now take another date seed and with a sharp knife and a good deal of force cut the stone across through the middle of the soft embryo which is now seen to be embedded in (Fig. 24, II, *e*), though readily detachable from, the hard horny material of the stone. Looking at the cross-section of the embryo with a pocket lens we see that it is not a homogeneous body but consists of recognizable parts neatly packed together: these include the first root, the tip of the stem, which has not yet begun to elongate, surrounded by a small tubular structure—the single seed-leaf or cotyledon. Stages in

SEEDS AND SEEDLINGS

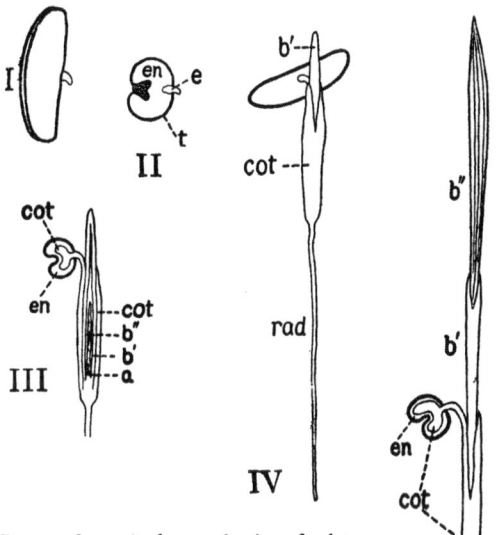

Fig. 24. Stages in the germination of a date seed. I. Side view of a seed from which the young root of the embryo is emerging. II. A section through the middle of the seed showing the embryo (*e*), the horny reserve food (*en*) and the thin brown seed-coat (*t*). III. Section of the seed and the young seedling: *cot*, the enlarged end of the seed-leaf (cotyledon) expanding as the reserve food (*en*) decreases; *cot* (on the right), the part of the cotyledon outside the seed; *a*, the apex of the future stem of the date palm; *b'*, *b''* the first foliage-leaves which have not yet emerged. IV. A later stage showing the elongated root (*rad*), the cotyledon-tube (*cot*) partially enclosing the first foliage leaf (*b'*). V. Section of the seed showing the enlarged cotyledon (*cot*) and the remnant of the reserve food (*en*). The seedling has now formed some lateral roots and the two first foliage-leaves (*b'* and *b''*) have become much longer. (After Thoday.)

the germination are shown in Fig. 24. A date, along with grasses and a host of other flowering plants, has only one seed-leaf, and on that account it was assigned by one of the first English botanists, John Ray (1627–1705), to a class which he called Monocotyledons (that is plants with one cotyledon). The other flowering plants such as the bean, castor oil plant and thousands of others, the embryos of which have two equal seed-leaves, he called Dicotyledons. In order to make out the structure of the horny material of the date seed we must examine a thin section of it under the microscope. We shall then find that it consists of cells which form an almost solid mass; the cell-cavities having been practically obliterated by the excessive thickening of the walls. This type of seed is relatively rare and affords an example of reserve food stored in unusually thick walls and not, as in most seeds, in the cell-cavities.

Before discussing further the nature of the reserve food which is always present in seeds, either outside or inside the embryo, we will describe as briefly as possible the changes accompanying germination. Seeds reach a certain stage of development while still attached to the parent-plant: when ripe they are set free in various ways, in many plants by a natural opening of the enclosing fruit along definite lines. In fruits which are fleshy and not dry there is as a rule no special opening mechanism: the seeds are often set free by animals eating the fleshy covering and passing the hard, indigestible seeds unharmed through the intestine. The relation of the seed to the fruit and the means by which seeds are released and dispersed is too wide a subject to be dealt with here: our main purpose is to convey a general idea of germination.

When fruits and seeds fall from the tree or smaller plant, the seeds do not usually germinate at once: the shedding of fruits, with their seeds, is followed by an interval during which the seeds remain in a dormant state until they germinate. The interval may be a few months, a few years, or possibly a hundred years. It has often been said that grains of wheat found in the wrappings of Egyptian mummies have been persuaded to germinate; but there is no satisfactory evidence of this. There are, however, many recorded instances of seeds germinating after lying dormant a hundred years or more. This ability of seeds to retain vitality, to remain alive though apparently dead and showing no external signs of life for long periods, is one of the great advantages conferred upon the higher or seed-bearing plants. During the quiescent period seeds may be carried some distance from the parent by wind, water, or animals, thus spreading the plant from place to place. Moreover, the high resisting power of seeds to adverse conditions is a safeguard against extinction of the species: the seeds survive long periods of drought or extremes of temperature which would be fatal to the plant as a whole.

The time arrives when a seed germinates; the dry and apparently dead body awakens to active life and the young plant gradually emerges. One of the essential factors leading to this reawakening is water: a dry seed can do nothing, the chemical changes required to convert the solid food-reserves into a form in which they can be absorbed, and thus used as fuel by the embryo, take place only in the presence of water: without water no growth is possible. Another necessary condition is a certain temperature: the plant machinery begins to work at a temperature which varies with different plants. Oxygen must also be available; that is to say seeds must have access to air. When these conditions are fulfilled the stage is set for the first act in the life-cycle. By the combined agency of the various factors the potential energy stored in the seed becomes converted into kinetic energy; pent up forces are released. The legacy of food left by the parent is the capital with which the young plant is able to start life and germination cannot begin until this capital becomes fluid. The first sign of germination is the appearance of a root which, as we previously said, at once grows downwards in response to gravity and makes the seedling fast in the ground. At a later stage the young stem stretches its cells, makes new ones and thus increases in length though in the opposite direction to the root. These and other early stages can be followed without difficulty when we watch the progress of a germinating bean (Fig. 25); it is unnecessary to describe them in full as they can be followed in the series *A–F* in Fig. 25. There is, however, one interesting point which deserves special mention: as a bean seedling develops we notice that the thick cotyledons never come out of the seed; they gradually shrivel and decrease in bulk within the seed-coat because the starch and other food stored in them have been gradually converted into sugar and other soluble substances, which are transferred to other parts of the young plant and provide it both with building material and the energy necessary for the movements of the root and stem during the unfolding of the first green leaves.

If we now follow the corresponding stages in the germination of a seed of the castor oil plant we find a different

SEEDS AND SEEDLINGS

state of affairs. Among other seeds illustrating this type of germination are the buttercup, shepherd's purse and many other common British plants. On cutting open a castor oil seed which has been softened and swollen by soaking in water we see that the inside of the seed is practically full

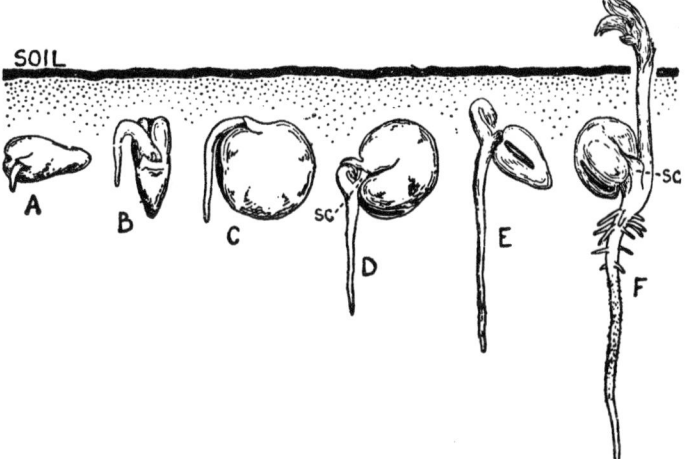

Fig. 25. Stages in the germination of a bean seed. *A*. The root coming out of the seed and growing vertically downwards irrespective of the position of the seed (*B* and *C*). In *D* the bent tip of the stem is seen, and at *sc* the stalk of one of the seed-leaves. In *E* the loop of the stem is free from the seed, and in *F* it has reached the light and is unfolding the first leaves. (After Yapp.)

of a light yellowish substance consisting of closely packed cells containing much oil (castor oil) and protein material: this food-store is outside the embryo. The embryo itself can be recognized lying in the middle of the oily food-store (Fig. 26, *A* and *B*): it consists of a very short spindle with one end bluntly pointed and easily seen, while the other end, which is very short and more difficult to see, is hidden by two extremely thin cotyledons lying in the mass of food. Though still small and very delicate the cotyledons already show a number of regularly disposed veins or conducting strands. When germination occurs the seed swells and ruptures the mottled and dark coloured coat; the root elongates and fixes in the ground the young axis or spindle of the seedling. Elongation of

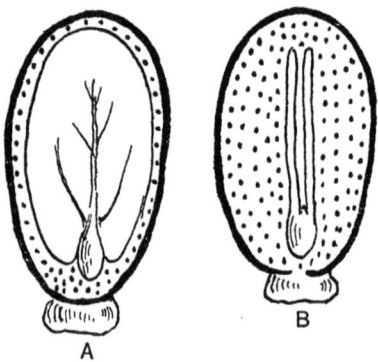

Fig. 26. Seeds of the castor oil plant cut into two showing (*A*) the embryo with one of the two cotyledons in surface-view lying in the oily store of food (dotted), and in side-view (*B*) both cotyledons in section, the short stem-apex and the larger root below.

the axis continues until after a time we see projecting above the surface of the ground a small stem bent into a loop, though its tip is not yet visible, but is still within the seed and attached to the pair of thin seed-leaves. What happens is this: the axis of the seedling increases rapidly in length and as both ends of it are fixed, the root-end in the ground and the tip of the stem-end still held within the seed, elongation of the intervening part causes the axis to bend into a loop. This method of emergence is clearly advantageous: if the stem-apex grew straight up its delicate tip would probably be broken by friction against the grains of sand and other hard particles in the soil, but by rising as a relatively strong loop this danger is avoided. Eventually as the loop becomes larger it exerts a stronger pull on the stem-apex with the cotyledons below it and these are dragged away from the seed-coat. The stem straightens itself vertically upwards and the two cotyledons expand into a horizontal position and begin to function as the first green leaves. In this type of seed the food-store becomes depleted as in the bean and all other seeds, but as it is outside the embryo the cotyledons which lie in it act as suckers or absorbing organs and transfer the food to the cells of the growing seedling.

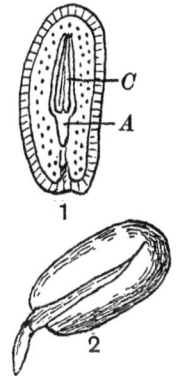

Another example in which the embryo lies in the middle of the food-store is afforded by the seeds of pine trees. The seeds of some pines have a fairly hard shell enclosing a kernel which consists of a tissue full of food, in the middle of which will be found the embryo consisting of a small cylindrical axis, the lower part being the root and the opposite end the apex of the future tree (Fig. 27). A circlet of long and narrow leaves surrounds the tip of the stem.

Fig. 27. 1. A pine seed cut through the middle. A woody seed-coat surrounds the food-store (dotted) in which lies the embryo. C, several cotyledons; A, the embryo root. 2. A germinating seed of a pine: the seed-coat has split and the root is growing into the ground.

It is interesting to trace the development of a pine seedling from the seed and to note the difference between the young foliage and the stiffer needles characteristic of the mature plant, which are not produced until the seedling has been growing for two or three years.

The germination of a date follows a rather different course: as in the bean and the castor oil plant the young root emerges first and carries with it both the tubular cotyledon, which rapidly increases in length, and the still undeveloped rudiment of the future stem. While the root and the cotyledon-tube are growing longer and penetrating deeper into the soil the tip of the tubular cotyledon remains embedded in the horny substance of the seed. In order to see what is happening within the seed we pull up date seedlings at different stages of their development and examine the seeds. We find that as the seedling increases in size the tip of the cotyledon grows larger and occupies more and more of the inside of the seed which is gradually becoming a soft mass (Fig. 24). This shows that the horny cellulose-walls have been destroyed; enzymes are produced and, as we have explained, they accelerate certain chemical changes which convert the hard horny food of the stone into soluble material which can be readily absorbed by the seedling. In the absence of the appropriate enzyme the breakdown of cellulose would be so very slow that years would pass before any detectable amount of change could be observed. At a later stage the first foliage leaves push their way through the cotyledonary tube at a place where there is an opening (Fig. 24, V) to admit of their upward passage to light and air.

These examples have not been described in detail because they can easily be studied more fully by anyone who cares to raise seedlings and follow their development. Attention has been called to a few of the more important points.

Reserve food occurs in many forms: starch is the commonest and most conspicuous because it occurs as relatively large grains. Proteins are very often found in seeds as granules of various size and form; many seeds contain oils and fats, and in a few, such as the date, the food is stored as cellulose in the thick walls of the cells.

The life of a green plant may be divided into three stages: the first begins with the fertilized egg, a single cell with a small, denser and well-defined, specialized portion of the protoplasmic contents known as the nucleus, which is made up in part of material belonging to the nucleus of the unfertilized egg and in part of nuclear substances given to it by the male germ. Within the nucleus of the fertilized egg are combined two sets of characters, one derived from the female parent, the other from the male. It is generally believed that the characters transmitted from generation to generation are represented by minute pieces of nuclear matter. A fertilized egg has within it enormous potentialities; given favourable conditions and the necessary supply of energy, it will develop into a complete plant resembling in all essential features the parents which provided the male and female reproductive cells. Within the protoplasm are concentrated latent forces which are destined to control and direct development along certain predetermined lines. As the young plant grows it becomes exposed to various external influences which together make up the environment, such as water, temperature, soil, and competition with other organisms. Its form and behaviour may be slightly altered or modified by these external forces; but in the main the adult may be described as an organism that has been gradually unfolded from an initial microscopic ball of protoplasm; an organism moulded by influences which are an expression of the dominating individuality of the fertilized egg.

As the egg grows the single cell divides into two, each division being preceded by the bipartition of the nucleus and an equal distribution of its component parts between the new cells. This process continues until from the egg is produced an embryo in which the root, stem, and first leaves of the future plant are recognizable as clearly defined members. These initial stages are rendered possible by food supplied directly by the parent plant to which the seed is still attached. The first stage in the life-history ends with the ripening of the seed, which contains the embryo and a store of food. There usually follows an interval after which in the presence of water and oxygen, growth is resumed and the energy of the stored food is used partly for building new tissues and in part for growth. The embryo becomes the seedling. In this period of its life the plant is still a dependent organism growing not actually on the parent but, by reason of the energy that is released through the agency of enzymes, from the capital inherited from the parent. The young seedling is at first unable to make its own living; its total weight in the early stages of development is not greater than that of the seed from which it comes. Finally, as soon as the seedling root is capable of active absorption of raw material from the soil, and as soon as the first leaves on exposure to sunlight produce the chlorophyll and so provide the mechanism for obtaining a supply of carbon first-hand from the air, the plant enters upon the third stage. Henceforward it grows as an independent individual which has broken contact with its parent and makes its own way in the world as a self-supporting and independent organism.

CHAPTER XI

Early Stages in the Evolution of Plants

So far we have drawn attention to the more striking differences and resemblances between the higher plants—in which category are included trees, herbs and other seed-producing plants since these are relatively complex in structure—and the higher animals. It is clear that, despite certain essential resemblances inherent in living things, the contrasts presented by the two organic kingdoms are more apparent than the features they have in common.

In order to appreciate the fundamental unity in origin of members of the plant and animal kingdoms we must examine in some detail certain microscopic organisms which have a very much simpler structure than the plants so far considered. Flowering plants begin life as single cells: as the fertilized egg of a tree grows into the larger embryo it is replaced by a mass of cells, which in the earlier stages of development are all alike and afford little evidence of division of labour, such as becomes increasingly apparent as growth proceeds and groups of cells acquire distinctive structural characters correlated with the shares they take in the co-operative activities of the whole plant-body. A tree, or indeed any of the higher plants, may in a general sense be compared with a well-ordered community in which the various kinds of work necessary for the general well-being are distributed among

EARLY STAGES IN EVOLUTION

groups of people fitted by their several qualities for the efficient performance of different shares in a co-operative undertaking.

All organisms begin life as single cells; the plants with which we are most familiar gradually become more and more complex in construction as development and cell-multiplication proceed. There are, however, very many organisms —both plants and animals—which remain single cells throughout life or at least until they reproduce other individuals like themselves. One such organism, a microscopic green cell known as *Euglena*, can often be found in stagnant water rich in organic matter. *Euglena* is a small piece of protoplasm without any encasing wall of cellulose: it is a naked cell capable of active movement in water. When in the motile state the minute body is seen to be longer than broad; it is blunt at one end and tapered towards the opposite, basal end (Fig. 28). To the blunt apex is attached a delicate thread—difficult to see—which by a rapid lashing movement from side to side propels the organism through the water much in the same way as the propeller drives an aeroplane through the air. When

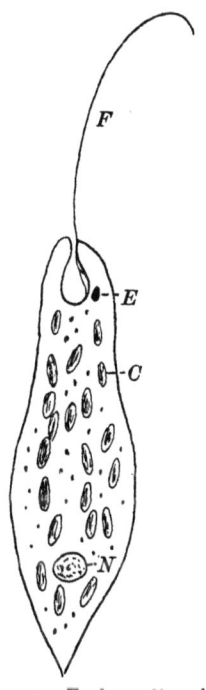

Fig. 28. *Euglena*. *E*, red eye-spot; *F*, flagellum; *C*, chlorophyll-bodies; *N*, nucleus. Very highly magnified.

at rest *Euglena* alters its form, loses its propeller, or flagellum (Lat. *flagellum*, a small whip), and becomes spherical. It is possible under a moderately high magnifying power to detect clear evidence of division of labour within the body or at least to see that the contents are not homogeneous or uniform in structure. We notice that the green colour (Fig. 28, *C*) is confined to a definite part of the living substance; we see also a small red spot (*E*) near the blunt end and, especially after killing the cell and staining it, a close inspection reveals the presence of a long flagellum (*F*). There are also other well-defined, recognizable pieces of material in the cell, for example a nucleus (*N*). The point is that even in this simple organism there is a distribution of offices, a division of labour: the organism is not by any means a mere fragment of uniform or undifferentiated protoplasm.

By means of the chlorophyll *Euglena* absorbs certain rays of light and, like an ordinary green plant, is able to utilize radiant energy and build up its food from air and water. It differs from most green plants in spending much of its life in active movement, and thus exhibits one of the more obvious attributes of animals. Another noteworthy peculiarity of *Euglena* is its ability to absorb ready-made nitrogenous and carbon compounds, a property that is characteristic of animals and not in any appreciable degree an attribute of green plants. We should not be far wrong if we called *Euglena* an organism which sometimes behaves as a typical animal and sometimes as a plant. It is one of many single-celled organisms which swing backwards and forwards between the animal and plant kingdoms. We cannot easily define its position and must be

EARLY STAGES IN EVOLUTION

content to regard it as an occupant of No Man's territory, a living thing which does not behave consistently either as an animal or a plant. Its method of reproduction is of the simplest kind and essentially primitive. When conditions are favourable reproduction is rapid: an individual increases in size by converting into its own living substance dead material absorbed from without, and then gradually splits lengthwise into two; the nucleus, the chloroplast, the red eye-spot all divide and finally two complete individuals are produced. The parent ceases to exist and is replaced by a pair of smaller *Euglenae*. *Euglena* shares with green plants the quality of irritability; it responds, for example, to the stimulus of light and the small red spot near the apical end is called the eye-spot because it is believed to be concerned in the perception of the rays of light. The response of *Euglena* to light can be demonstrated in the following way:

Take water from a stagnant pool where *Euglena* is abundant and put some of it into a glass basin covered, except at one place where a clear piece of glass is left, with black paper: expose the vessel to a source of light opposite the uncovered part of the glass. After a short time the *Euglenae* congregate in that part of the water which is illuminated: with the naked eye, if the organisms are present in quantity, one can see a green patch where the light has penetrated.

From what has been said it would seem that in an organism such as *Euglena* the distinguishing features which enable us to recognize members of the plant and animal kingdoms are inapplicable. This suggests the possibility of a common origin of plants and animals from ancestors occupying a primitive borderland territory and which had not progressed far enough in evolution to acquire the features now regarded by us "afterthoughts of creation" as distinctive of the two kingdoms. We may next enquire whether there are in existence other simple organisms which, though comparable in structure with *Euglena*, are more plant-like and exhibit more clearly the characteristics of green plants, organisms which have passed beyond the relatively primitive stage illustrated by *Euglena* and similar members of the borderland territory, and may fairly be regarded as simple representatives of the vegetable-kingdom.

The organism which we will now briefly describe is usually assigned to the plant-kingdom though zoologists do not definitely exclude it from their province: it is called *Chlamydomonas* and like *Euglena* lives in water though it prefers fresh to stagnant water. *Chlamydomonas* differs from *Euglena* in having a cell-wall composed of cellulose. It is broadly oval in form and consists of a single cell: at one pole, the apical end, there are two equal flagella or cilia as they are usually called (Lat. *cilium*, a hair) which act as propellers (Fig. 29, *a*), though they are often thrown off and the cell remains for a time stationary. In its internal structure *Chlamydomonas* resembles *Euglena*: it has a chloroplast, a red eye-spot, a nucleus and other specialized organs. The possession of an elastic and permeable cell-wall is essentially a plant character, and another more important feature, in which it resembles all green plants, is its method of nutrition: the cell is self-supporting and does not normally make use of ready-made carbonaceous and nitrogenous food.

EARLY STAGES IN EVOLUTION

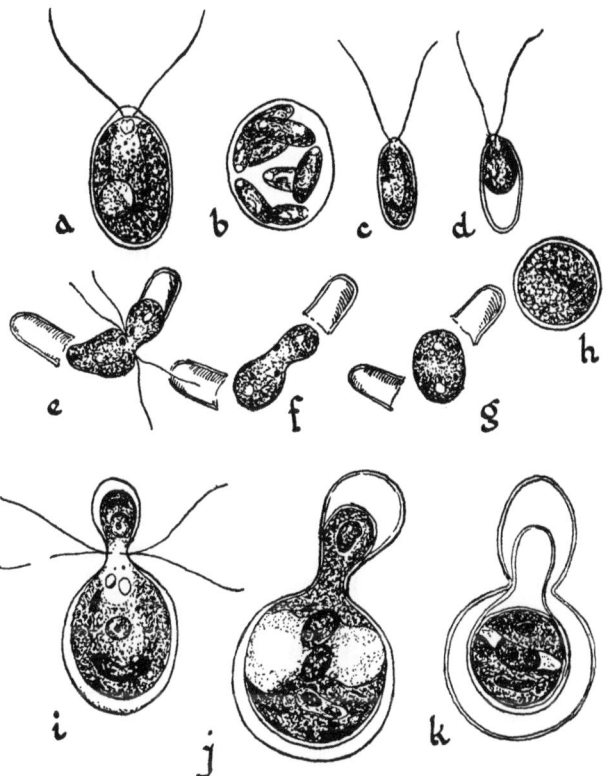

Fig. 29. *Chlamydomonas*. *a*, a complete individual; *b*, the plant has produced by successive divisions eight reproductive cells which are still enclosed by the wall of the parent; *c* and *d*, a single sexual reproductive cell after liberation; *e–g*, two sexual cells in different stages of coalescence to form a single spherical cell, *h*: note the discarded cell-walls which originally encased the protoplasm; *i–k*, stages in the union of two sexual cells of another *Chlamydomonas* in which the inequality in size is greater. (From Godwin's *Plant Biology*.) Greatly enlarged.

The method of reproduction is less primitive than in *Euglena* and throws light on the methods and course of evolution. *Chlamydomonas* reproduces itself in different ways, two of which will be described. Experiments indicate that the adoption of one or other method of reproduction is determined by the nature of the environment, that is the amount of food material available and other factors. The first and simplest method consists of an equal division of the protoplasm and other contents of a complete individual into two parts: the cell-wall of the parent remains unaffected by these changes in the enclosed material and it now surrounds two separate entities, each fully equipped with all the essential parts. The two new individuals divide in the same way as the parent cell and the original *Chlamydomonas* is replaced by four equal and smaller naked cells still lying loose within the old wall. Eventually the wall is ruptured; the products of the successive division are set free and, after increasing in volume by incorporating into their substance material from the surrounding water and the air it contains, they clothe themselves with a cellulose membrane formed by the protoplasm. Each of the two parts, or daughter-cells, is provided with a couple of propellers (cilia). Thus four, or it may be as many as eight, new *Chlamydomonas* individuals are produced by repeated bipartition of one parent.

In the second method of reproduction the same process of division is followed but the subdivision of the contents is carried further until there are eight (Fig. 29 *b*), sixteen or thirty-two pieces, each a complete individual on a small scale, usually naked though occasionally provided with a wall. On the disorganization of the original wall the separate units gain their freedom and swim about in the water by means of the rapidly moving cilia (Fig. 29, *e–g*).

Let us now note the distinguishing features of this second method: the separate units do not, as in the first process, grow directly into full-sized cells, but as they move about pairs come together and gradually coalesce to form single naked cells (Fig. 29, *h*) which proceed to grow larger and clothe themselves with walls. Each new *Chlamydomonas* is therefore made of two separate immature individuals which have completely fused. If, as often happens, the two pieces of protoplasm, with chlorophyll, nucleus, and other constituents which coalesce, are derived from two separate parents the new individual contains a mixture of the substance of two *Chlamydomonas* cells and inherits two sets of characters. Here we have an example of reproduction by the union or conjugation of two cells from separate parents, reminding us of what is called sexual reproduction in plants and animals. There is, however, a difference: in *Chlamydomonas* the two units which fuse are exactly alike, so far at least as we can see: the terms male and female seem to be inappropriate to a union of this kind since we cannot say which of the two conjugating cells has the properties of a female or of a male. It is reasonable to regard this type of reproduction as one of the earliest steps leading to a true sexual method. *Euglena* has no such method of reproduction; it reproduces its kind only by simple division of the body into halves.

There are several different kinds, or species of *Chlamydomonas*: in most of them conjugation occurs between pairs of equal cells alike in all respects. In at least one species it has been found that in some individuals the process of division of the contents is carried further than in others with the result that two sizes of reproductive cells are produced. From one *Chlamydomonas* a large number of relatively small motile pieces escape and from another parent of the same species a smaller number of larger pieces are set free. Both large and small swim about in the water: the larger ones seem to be less active and come to rest sooner than the more active smaller cells. In the majority of species these motile sexual cells conjugate in pairs, but in the species we are now describing union occurs between unequal pairs, always between a large and more passive cell and a smaller and more active cell (Fig. 29, *i–k*). Here we have not only fusion of two separate units, which are reproductive cells unequal in size and differing in activity; one is more passive than the other. From our knowledge of sexual reproduction in the higher plants and animals, where the female cell is more passive and larger than the male, we are justified in describing this type of *Chlamydomonas* as a plant exhibiting a simple form of bisexuality.

This microscopic water-plant affords a striking illustration of some of the first steps in the evolution of more complex forms: the reproduction of *Chlamydomonas* is sometimes purely asexual, that is a new individual is developed without any previous act of conjugation, without a mixture of protoplasm, as in *Euglena*. The next step is seen in the coalescence of two separate reproductive cells either from the same parent or—a higher type—from two parents. This may be considered the most primitive and simplest kind of sexual reproduction though as yet the sexes are not clearly defined: there is no recognizable maleness or femaleness. A more advanced type is that in

EARLY STAGES IN EVOLUTION

which two parents are involved and the conjugating cells differ from one another in size and behaviour.

We may now discuss in a few words the bearing of the facts so far recorded on the great problem of evolution. The existence of *Euglena* and similar organisms, which in their manner of feeding conform to the characteristic behaviour of both animals and green plants, furnishes a strong argument in support of the view that plants and animals are derived from common ancestors. If the animal and plant kingdoms are branches from one primitive stock, the contrasts between them, which are so obvious when we compare the more advanced examples of each, must be an expression of gradual divergence of two lines of evolution from one starting-point.

There was no doubt a time in the history of the world when the organic kingdom was represented only by inconceivably small specks of living protoplasm; we can only speculate on the mystery of the birth of life, picturing, it may be, a lifeless ocean containing, in solution, mineral matter washed from the rocks of the earth's surface, an ocean warmed by the sun; the creation at some early period of chlorophyll which provided living cells with the power of seizing radiant energy and converting it into stores of potential energy represented by the organic substances made in a plant's body. How and when the dead became living we do not know; the chemical elements which enter into the substance of protoplasm were present as simple salts or some of them in a free state in the seawater. There must have been a creative act by which lifeless matter became instinct with the properties we call life. The difference between matter that is dead and matter that is alive cannot be adequately defined. We cannot speak of the one as motionless and inert and the other as endowed with motion, and active: all matter is in constant movement. The birth of protoplasm marked the beginning of an entirely new world; a world which was wholly inorganic received a germ destined to evolve into a world that was organic. Each of the two worlds became indissolubly connected one with the other, both obeying the same set of laws and both in consort playing their parts in the great drama of evolution. Once the barrier between the dead and the living was passed the germs of life increased rapidly in size; the invisible became visible. By slow degrees the protoplasmic particles reached a size which brought them within the range of human vision aided by the highest magnifying lenses; in them, small as they are, we can detect definite evidence of specialization in the presence of minute entities differing from one another in appearance, each taking a definite share in the living mechanism. Bacteria, organisms such as *Euglena*, and the still more elaborate *Chlamydomonas* are examples of some of the simpler products of evolution which we can submit to intensive study, though they are very far removed from the most primitive progenitors of the organic world.

Let us now see whether we can discover any indication of later stages in progressive evolution by which single-celled organisms such as *Chlamydomonas* may have produced descendants with bodies consisting of several cells. One of the simplest plants which bears unmistakable signs of a close relationship with *Chlamydomonas* is a microscopic, green freshwater form known as *Gonium*.

EARLY STAGES IN EVOLUTION

A *Gonium* can often be found in freshwater ditches and ponds: it consists of sixteen cells arranged in the form of a square (Fig. 30). Each cell is exactly like its neighbours and all of them closely resemble a complete *Chlamydomonas*.

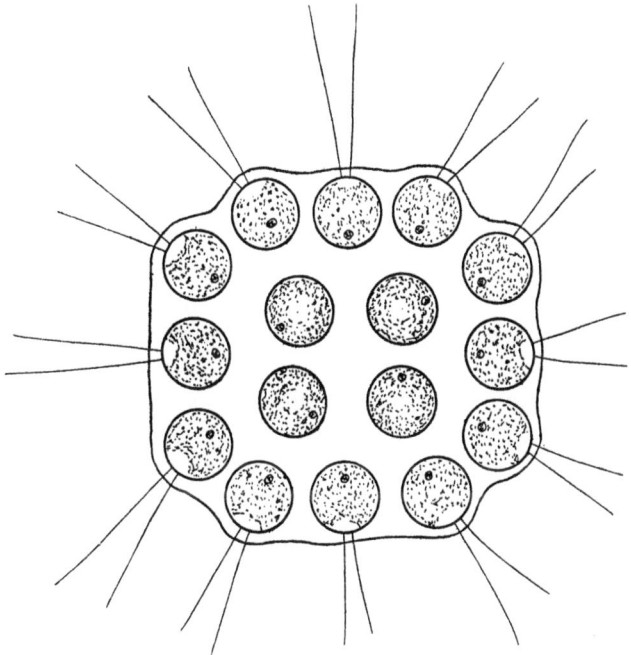

Fig. 30. Diagrammatic drawing of a *Gonium*. Greatly enlarged.

The sixteen cells are all provided with two equal cilia and the whole group moves through the water with an erratic gait: the movement of a *Gonium* has been compared with that of a dinner plate as it turns over and over when sinking in the water of a swimming bath. The plant differs from *Chlamydomonas* in being many-celled, though the union of cell with cell is not very intimate: each cell retains its individuality in a greater degree than the cells which compose the tissues of higher plants. We speak of *Gonium* as a colony of identical cells; it is a colonial form rather than a plant composed of a true tissue. Its origin from a simple-celled ancestor can easily be imagined: we know that *Chlamydomonas* reproduces itself by the division of its body into smaller pieces each endowed with the parental characters. If the progeny of small cells instead of swimming about independently grouped themselves into a regular pattern and in a constant number, each unit being united with its associates by delicate threads of protoplasm, and the whole colony invested with a slimy material formed from the walls of the several cells, the result would be a plant of the *Gonium* type. In *Gonium* and other similar plants every cell does the same work, each participates equally in the communal life; there is no division of labour among the component cells. Growth and reproduction are included in the work which each cell performs, whereas in most plants these two functions are allotted to different kinds of cells.

CHAPTER XII

Later Stages in Evolution

A short account of a still larger and more elaborate colonial type of plant will serve to illustrate a further step in evolution. There is occasionally found in ditches and ponds a plant known as *Volvox* which is just large enough to be seen with the naked eye as a green sphere rolling about in the water. The colony or plant-body is made up of several hundred or even several thousand cells arranged side by side in a single layer which forms the wall of a relatively large hollow sphere (Fig. 31, *A*). The cells of the wall in a young *Volvox* are all alike but in a mature individual one notices that a few are rather larger than the others. Each cell, like the cells of a *Gonium*, is in essentials similar to a complete *Chlamydomonas*, having a pair of cilia, a nucleus, a red eye-spot and other recognizable parts. The cells of *Volvox* are more closely united than in *Gonium*. The larger cells which occur here and there in the spherical layer of a *Volvox* are concerned with reproduction and, as in *Chlamydomonas* and *Gonium*, the plant reproduces itself both asexually and sexually. The simpler method of reproduction consists in the repeated division and growth of the few larger cells, often about ten in number, until from each is produced a new *Volvox*: this becomes detached from the bounding wall of the colony and rolls about in the central cavity of the sphere until such time as the parent plant is ruptured, when it escapes as an independent organism. Sometimes *Volvox* resorts to a much more elaborate method of reproduction: certain cells in the wall of the sphere grow larger than their neighbours; some eventually produce, by the repeated cutting up of

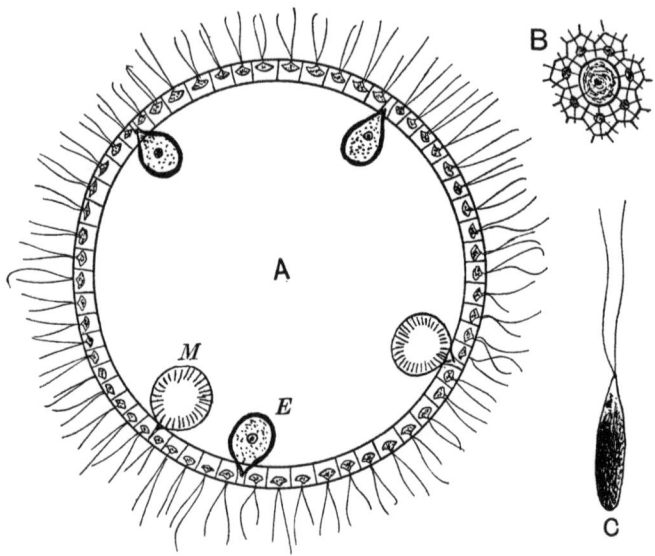

Fig. 31. *Volvox*. *A*. Diagrammatic drawing of a section through the spherical body of a *Volvox*, on a much enlarged scale, showing the wall composed of numerous cells, each with two cilia; also some cells, *M* and *E*, larger than the great majority of the wall-cells. *E*, the egg-cells; *M*, two groups of male cells. *B*. Surface-view of a small portion of the spherical plant, showing a single egg-cell in the centre surrounded by several sterile cells. *C*. A single male cell enlarged several hundred times. (After Janet.)

their living body, a large number of minute, long and narrow pieces, each consisting of a relatively large nucleus (derived from the parent nucleus), a long whip-like flagellum, and other organs. These cells are liberated by the bursting of the retaining wall and swim about actively. They are the male reproductive cells (Fig. 31, C). A few other cells of the sphere also increase in size, but they remain undivided, simple units with no power of movement. These passive cells, which are the female reproductive bodies, or eggs (Fig. 31, E), differ enormously in size from the small male cells. Many unnecessary details are omitted from this description. The important fact is that an egg is eventually fertilized by one of the motile male germs; the essential feature of this act of union is the fusion of the two nuclei. Stimulated by fertilization the egg proceeds to grow and gradually by repeated division is transformed into a new *Volvox*.

The main points are these: *Volvox* is made up of numerous cells—it is a multicellular organism; most of the cells are identical, but a few in each plant acquire distinctive characters and take over the work of reproduction, which is asexual or sexual according to circumstances. There is a very definite division of labour within the colony; the greater part of each individual plant is concerned only with nutrition and growth, a much smaller part is responsible for reproduction. The cells which form what we may call the sterile portion of the plant, the purely vegetative and non-productive cells, have a comparatively short life since each brood of young *Volvox* spheres obtains its freedom through the destruction of the parent. On the other hand, the reproductive cells, both male and female, are participants in the production of a continuous series of generations. It is true that a male cell does not persist as such for more than a very short time, but those which fuse with the female cells persist in part by their nuclear material, which may be regarded as the carrier of the qualities transmitted by the male parent. The nucleus of the male cell becomes merged into the nucleus of the egg. This is exactly what happens in even the highest plants and animals: the body is mortal, the germ cells are in comparison immortal. We see therefore that *Volvox*, despite its small size, agrees in certain essential features with plants far removed from it in diversity of construction. It is, moreover, interesting and surprising to find that in so simple a plant as *Volvox* the difference between male and female reproductive cells is as well marked as it is in the higher animals and plants.

By comparing the structure and the life-histories of a number of plants having in common certain features which are indicative of relationship, it is possible to gain some insight into the nature of the transition in the course of evolution from one type to another. It is reasonable to suppose that *Chlamydomonas*, *Gonium* and *Volvox* represent closely related forms in an ascending series. No effort of the imagination is needed to picture the development of a plant like *Gonium* from single-celled ancestors comparable with *Chlamydomonas*, and though the step is a higher one, we can well believe that *Volvox* is derived from simpler ancestors comparable with *Gonium*. There are indeed certain freshwater plants which in structure, and in the allocation of functions to different cells, occupy positions intermediate between *Gonium* and *Volvox*. The types chosen for description are, however, adequate for our present purpose as illustrations of some of the evidence on which views on evolution are based.

It is obviously impossible in a short space to give even

the bare outlines of the numerous problems with which we are confronted when an attempt is made to trace the origin of one set of plants from another, or to follow step by step the evolution of various plant-forms that bear clear evidence of descent from a common ancestral stock. The examples briefly described in the foregoing account are of special interest because they throw light on the nature of some of the stages in the progressive elaboration of structure. They illustrate a gradual advance in closely related plants from the simplest method of reproduction, in which the parent divides into two, to the slightly higher form of reproduction in which special cells, produced by repeated divisions of the body of the parent, are capable without union with other cells of growing into new individuals, and finally to the still more advanced method characterized by the production of two kinds of reproductive cells, male and female.

It may, however, be said and with truth that the few examples passed in review do not enable us to form any clear conception of the methods by which the higher plants, such as trees and other flowering plants, have been evolved from simple forms such as *Volvox*. There seems to be very little in common between members of the vegetable kingdom so far apart as the small, actively motile plants we have considered and an oak tree. *Volvox*, for example, is composed of a few thousand cells most of which are alike in size and structure as also in the shares they take in the life of the organism: a tree is built up of countless millions of cells forming many different kinds of tissues, each tissue characterized by certain structural features and taking its own special share in the life of the plant. Moreover, in *Volvox* the body is relatively homogeneous; it is not subdivided into members corresponding to the roots, stems and leaves of the higher plants. None the less the differences, great as they are, are differences in degree rather than in kind.

There are, no doubt, many people who would still describe a botanist as a person who collects and presses plants and is expected to be able to name at sight any specimen submitted to him. The science of botany includes many branches and it is impossible for the average student thoroughly to explore any one of them. It may, however, be said that despite modern specialization consequent on the advance of knowledge, every botanist takes an interest in evolution: whatever his special line of research may be, the general question of development of the plant-kingdom makes a strong appeal to him. Since Darwin's *Origin of Species* was published rather more than seventy years ago, we have learnt much both from plants that still exist and from plants that have long been extinct: none the less it is hardly an exaggeration to say that there has been comparatively little advance towards an agreed view on the relationship of the various classes and groups of plants and their respective lines of evolution. It is generally admitted that there must have been a gradual development of the more specialized and more elaborately constructed plants from less complex ancestors. Botanists are more conscious of the difficulties of the enquiry into the precise methods of evolution than they were seventy years ago.

One of the most puzzling questions is the relation to

LATER STAGES IN EVOLUTION

one another of the larger divisions of the plant-kingdom: it used to be a common practice to represent the various groups as so many different branches of one genealogical tree, thus implying a direct relationship between one group and another. In recent years the tendency has been to represent the progress of development during the history of the earth as a series of more or less independent lines, rather than as a number of divergent branches meeting in one common stem. It is probably more in accordance with fact to represent evolution as many disconnected lines each starting from a separate source, than to think of it in terms of a single tree with many branches from one original trunk.

The main object of the present chapter is to show by the description of a few simple plants and the still simpler organism *Euglena* that the statement made in the first chapter, namely that the plant and the animal kingdom had a common origin, rests on a substantial foundation.

We have touched only the fringe of the curtain which hides the mystery of life. Words written by Seneca rather more than nineteen centuries ago are still true:

> We imagine that we are initiated into the mysteries of nature, but we are still hanging about her outer courts.

The whole panorama of the world as we see it, with its teeming multitudes of plants and animals, the rocky crust rising into folded mountain-chains, covered with the sand of deserts or hidden by tropical forests, with its rivers, seas, and oceans, is but one scene in a drama extending over hundreds of millions of years. In the crust of the earth are entombed fragmentary remains of a mere fraction of the hosts which occupied the stage in the successive ages of the world, each stage having its own scenery differing in a greater or less degree from that of the age in which we live. The ambition of the evolutionist is to reconstruct within the narrow limits that are possible the development of plant-life, a development by no means uniformly progressive but frequently retrogressive. The province of the student of evolution is the whole world, the world as it is and as it was: his aim can never be more than partially achieved, he can never hope to reach his goal; but he can derive pleasure from his search for origins which will enable him to see the present face of the earth with an understanding and an enchantment intensified and illumined by a knowledge of the past.

It would be difficult to select a more fitting conclusion to this brief reference to the unsolved problem of evolution than the words written by Charles Darwin at the end of the last chapter of the *Origin of Species*:

> There is grandeur in this view of life, with its several powers, having been originally breathed by the Creator into a few forms or into one; and that, while this planet has gone cycling on according to the fixed law of gravity, from so simple a beginning endless forms most beautiful and most wonderful have been, and are being evolved.

BOOKS SUGGESTED FOR FURTHER READING

The following selection is made from a large number of books several of which might confidently be recommended as alternatives to those mentioned. It would be easier and less invidious to compile a longer list, though that would probably be less helpful.

The books numbered 3, 4 and 8 are not specially designed to meet the requirements of examination schedules. Books 1, 2, 5, 6, 7 and 9 may be described as more definitely school books suitable also for elementary university students. Books 10 to 13 are larger and more comprehensive text-books. Books 14 to 17 belong to a different category and are what botanists call "Floras": they are guides to the identification and characters of plants represented in the British flora.

1. *Botany*: a junior book for schools. R. H. Yapp (Cambridge University Press).

    "Intended to provide a sound course of instruction in the fundamental principles of Botany."

2. *Botany*: a text-book for senior students. D. Thoday (Cambridge University Press).

    "Intended primarily for use in connexion with the Senior Cambridge Local Examinations."

3. *An Introduction to Structural Botany.* Part I. Flowering Plants. Part II. Flowerless Plants. D. H. Scott (A. and C. Black, London).

    "A first guide to the study of the structure of plants."

4. *Elementary Botany.* P. Groom (G. Bell and Sons, London).

    "Students of plant-life *must* look at plants, and this book is specially designed for use during the process."

5. *Plant Biology*: an outline of the principles underlying plant activity and structure. H. Godwin (Cambridge University Press).

   "This book lays emphasis on the physiological point of view and on consideration of the simpler characters of the physico-chemical background of plant-life."

6. *Elements of Plant Biology*. A. G. Tansley (A. and C. Black, London).

   "Intended primarily for medical students and others who do not necessarily intend to continue the study of Botany."

7. *School Botany*. Macgregor Skene (Oxford University Press).

   "An introduction to botany through the study of the structure and functions of the flowering plant."

8. *Life of Plants*. Sir Frederick Keeble (Oxford University Press).

   The author's object is "to suggest that Science is more than doctrine —an illumination of life".

9. *An Introduction to Plant Physiology*. W. O. James (Oxford University Press).

   For senior school or junior university students. The author "seeks to give a balanced account of the more elementary aspects of plant physiology".

10. *An Introduction to the Study of Plants*. F. E. Fritch and E. J. Salisbury (G. Bell and Sons, London).

    "An elementary introduction to the manifold aspects of plant-life."

11. *An Introduction to the Structure and Reproduction of Plants*. F. E. Fritch and E. J. Salisbury (G. Bell and Sons, London).

    A sequel to No. 10.

12. *Botany of the Living Plant*. F. O. Bower (Macmillan and Co. London).

    Framed on the lines of the annual course of elementary lectures given by the author in Glasgow University.

13. *A Text-book of Botany*. R. Strasburger and other German Professors. A translation. (Macmillan and Co. London).

    A more comprehensive book.

14. *A School Flora* for the use of elementary botanical classes. W. Marshall Watts (Longmans, Green and Co. London).

15. i. *Handbook of the British Flora.* G. Bentham (L. Reeve and Co. London).
  ii. *Illustrations of the British Flora.* Drawings by W. H. Fitch and W. G. Smith.
     A companion volume to the *Handbook.*
  iii. *Further Illustrations of British Plants.* R. W. Butcher and F. I. Strudwick.
16. *The Student's Flora of the British Isles.* Sir J. D. Hooker (Macmillan and Co. London).
17. *Manual of British Botany.* C. C. Babington; 10th edit. edited by A. J. Wilmott (Gurney and Jackson, London).

GLOSSARY

Atom (Gk. *atomos*, indivisible). The smallest particle of an element which can exist in combination with other similar particles of the same or of different elements: the smallest quantity of matter possessing the qualities of a particular element.

Calorie (Lat. *calor*, heat). The unit of heat required to raise the temperature of 1 gram of water through 1° centigrade.

Cambium (Low Lat. *cambio*, change, i.e. barter). A hollow cylinder or strip of formative tissue which manufactures wood and other tissues by repeated division of its cells. It is a tissue which causes change or exchange by producing new cells which become different from the parent-cells. The stem of a tree increases in girth as the result of cambial activity.

Capillarity (Lat. *capillus*, a hair). The rise of water in a tube with a very narrow bore which only admits a hair: the forces concerned are the force of cohesion between molecules of the water and the force of adhesion between the water and the glass of the tube. Water rises in a capillary tube to a level above that of the surface of the water in which the end of the tube is inserted.

Carbohydrate. Compounds, such as sugar, starch, etc., containing carbon, hydrogen and oxygen, the two latter being present in the same proportion as in water, that is two atoms of hydrogen and one atom of oxygen (H_2O).

Carpel (Gk. *karpos*, fruit). The female organs of a flower situated in the centre either above the insertion of the petals, e.g. in the buttercup where there are several separate carpels, or below the other parts of the flower, e.g. in the daffodil where the three carpels are united to form the green swelling between the flower and its stalk. The female part of a flower, i.e. the carpel or carpels, is often called the pistil. The carpels are closed cases containing within the internal cavity (ovary) the ovules which after fertilization form seeds, the carpel growing into a fruit or contributing to the fruit. On the carpel is a small projection, the stigma, often supported on a stalk, the style, which receives the pollen.

Cell (Lat. *cella*, a small apartment). The name cell originally given to the dead membrane of a single compartment in cork-tissue is now used for what may be called the unit of structure, a small piece of protoplasm, with its nucleus, which may be naked or surrounded by a cell-wall.

Cellulose. The name given to certain complex carbohydrates some of which form the chief constituent of the walls of most plant-cells. Cotton is composed of the cellulose walls of hairs: cellulose is a valuable vegetable product used in the manufacture of artificial silk, paper, etc.

Chlorophyll (Gk. *chloros*, grass-green; *phullon*, a leaf). The substance which gives the green colour to plants: it has a very complex composition and is a mixture of four pigments.

GLOSSARY

Chloroplast (Gk. *chloros*, grass-green; *plastos*, moulded or formed). A well-defined piece of protoplasm which holds in its substance the chlorophyll.

Cilia (Lat. plural of *cilium*, an eyelid, eyelash). A cilium is a very slender thread, visible only on high magnification, which occurs either singly, in pairs, or in large numbers on certain plant-cells and by rapid movement propels them through water.

Cotyledon (Gk. *cotuledon*, a cup-shaped cavity). The first leaves developed on an embryo-plant while it is still embedded in the seed. Some embryos, e.g. an embryo pine, bear several cotyledons; others have two, e.g. the flowering plants known as Dicotyledons, including the majority of trees and herbs; or one, e.g. grasses, palms, and other Monocotyledons.

Element. Substances which have not hitherto yielded on analysis other substances: the simple chemically indivisible substances of which all matter consists. Certain radioactive elements which break up into material unlike themselves are exceptions. There are about ninety known elements, many of which do not exist in nature in the free state.

Enzymes (Gk. *en Zume*, in yeast). Substances produced by animals and plants, which were formerly called ferments; they are endowed with the power of accelerating various chemical reactions without themselves entering into the composition of the final products. Examples of enzymes are: the enzyme of yeast which brings about the formation of alcohol and carbon dioxide from sugar; the enzyme (diastase) which converts starch into sugar. Enzymes, a very large number of which are known, effect changes which could be made in the laboratory only with powerful reagents or at a high temperature.

Epidermis (Gk. *epi*, upon; *derma*, skin). The surface-layer of cells on leaves, and on all stems in which it has not been replaced by cork.

Eye-Spot. The small red spot near the apical end of many simple organisms in which the whole individual is a single cell and in the cells of slightly less simple organisms composed of a colony of cells. The eye-spot is said to be sensitive to light.

Flagellum (Lat. *flagellum*, a small whip). A delicate hair-like thread which by its rapid movements propels a free-swimming cell through the water. Cf. Cilia.

Geotropism (Gk. *ge*, earth; *tropos*, direction). Curvatures executed by plant-organs in response to the action of gravity. The downward growth of a root is an example of positive geotropism; the upward growth of a stem away from the centre of the earth is spoken of as negative geotropism.

Heliotropism (Gk. *helios*, the sun; *tropos*, direction). Curvatures of plant-organs in response to the stimulus of light.

Kinetic Energy (Gk. *kinetikos*, moving). The energy of motion. A falling weight illustrates kinetic energy: when a complex substance is broken up into simpler substances the stored up, or potential energy, is rendered available as energy in a mobile state, that is kinetic energy.

GLOSSARY

Molecule (Lat. *molecula*, diminutive of *moles*, a small mass). The smallest particles of matter which exist as separate entities, not necessarily united with other molecules; in contrast to atoms (q.v.).

Nucleus (Lat. *nucleus*, a kernel). A specialized portion of the protoplasm of a cell characterized by its comparative density: it plays an essential part in reproduction and in the transmission of hereditary characters.

Osmosis (Gk. *osmos*, thrust, push). When a substance, e.g. sugar, is dissolved in a solvent, e.g. water, and the two are separated by a membrane, which is impermeable to sugar and permeable to water, the water passes into the sugar solution by osmosis. The sugar solution behaves as if the sugar were exerting pressure, which is spoken of as osmotic pressure.

Ovary (*ovarium*: formed from Lat. *ovum*, egg). The chamber containing the ovules; an essential part of the female organ (pistil) of a flower. See Carpel.

Ovule (Lat. *ovulum*, diminutive of *ovum*, egg). The small bodies within the ovary of the carpel (q.v.) which become seeds.

Oxide (Fr. *oxy-gène*, and *-ide* after the ending of the French *acide*). A compound formed by the combination of the gas oxygen with certain other elements, e.g. water is an oxide of hydrogen; carbon dioxide gas (CO_2) is an oxide of carbon.

Phloem (Gk. *phloios*, bark). That part of the conducting tissue which includes the sieve-tubes (q.v.) and conveys the manufactured food to all regions of the plant where it is being used for cell-formation and growth. Phloem is the inner part of the bark of a tree: it occurs in all plants with well-developed conducting tissue, ferns, etc., cone-bearing trees, and flowering plants.

Potential Energy. Energy of position as distinct from energy of motion (kinetic energy, q.v.).

Proteins (Gk. *prot*—, first, with suffix). Complex nitrogenous substances containing in addition to the elements carbon, hydrogen and oxygen (which compose carbohydrates), nitrogen and often phosphorus and sulphur. Proteins are characteristically products of living cells: the white of egg (albumen) is one of a very large number of proteins.

Protoplasm (Gk. *protos*, first; *plasma*, moulded or formed). The highly complex viscid substance, largely proteins, of which all living cells and tissues consist.

Salts. Compounds formed from an acid by the substitution of a metal for the element hydrogen.

Seeds. Product of sexual reproduction, distinctive of the higher plants, e.g. cone-bearing trees and flowering plants. Seeds are formed from ovules (q.v.) and contain an embryo with a store of food.

Sieve-Tube. Tubular cells much longer than broad with small areas on their walls penetrated, like sieves, by pores. Their chief function is the conveyance of diffusible organic material from one part of the plant to another; they constitute an essential part of the conducting tissue, external to the wood of a tree, which is known as phloem or bast.

Spore (Lat. *spora*, sowing, seed). An asexual reproductive cell which grows directly into a new plant or, in all but the plants belonging to the lowest division of the vegetable kingdom, into a stage in the life-history characterized by the production of male and female organs.

Stimulus (Lat. *stimulus*, goad). An influence which produces a change in behaviour: different stimuli induce different reactions, e.g. gravity, light, etc. Cf. geotropism and heliotropism.

Stoma, plural **Stomata** (Gk. *stoma*, mouth). The name stoma is applied to the minute openings on the surface of a leaf or a young green twig. A stoma consists of two cells known as guard-cells which enclose between them a pore that is closed or opened by their contraction and expansion. By means of the stomata the internal tissues of a plant are brought into direct relation with the outer air.

Tissue (Old Fr. *tissu*, applied to a rich kind of material). The fabric of which a plant-body is composed: the name is applied to groups of cells of similar structure performing a common function. The plant-body is made up of different tissues or tissue-systems to which different names are given indicative of their contributory shares to the life of the whole: e.g. mechanical tissue characterized by the great strength of the cells, conducting tissue consisting mainly of long tubes well suited as channels for the transport of water and food.

Transpiration (Lat. *trans*, through, and *spirare*, to breathe). The giving off of water-vapour from green leaves and twigs through the stomata (q.v.).

Tropism (Gk. *tropeo*, I turn). Various movements or curvatures produced in plant organs by stimuli having relation to the direction of the stimulus.

Vessel (Lat. *vascellum*, a small vase or urn). Dead tubular structures derived from vertical rows of cells (see p. 45) with walls in which the original cellulose has been largely converted into a substance known as lignin (Lat. *lignum*, wood); one of the chief functions of vessels is the conduction of water.

INDEX

Air, composition of, 27
Alcohol, conversion of sugar into, 71
Ammonia, 87, 88, 91
Animal kingdom, common origin of plant and, 2, 43, 110, 113, 119, 129
Animals, comparison of plants and, 1–8
— their inferiority to plants, 22
Ascent of sap, 77–82
Asexual reproduction, 118
Ash of plants, 33, 34
Atmosphere, composition of the, 27
— pressure of the, 77
Atom, 24
Azote, 27

Bacillus, 86
Bacteria, 42, 84–91
Bacterium, 86
Bean, germination of seed of, 102, 103
Beet-root, red cell-sap of, 75
Birch tree, stomata in leaves of, 55, 56
Bleeding of stems, 76
Breathing of animals and plants, 33, 68; *see also* Respiration
Brosse, Guy de, 1
Browne, Sir Thomas, 2
Buckland, W., 83
Buttercup, flower of, 95

Calories, 32, 33
Cambium, 7, 50
Capillarity, 78
Carbohydrate, 29–31
Carbohydrates, manufacture in the plant of, 31, 64–72
Carbon, green plants and, 28–31, 35
— proof of its presence in air, 28
Carbon compounds, 29, 50
Carbon-cycle, 92
Carbon dioxide, 26–29, 33, 87
Carpel, 95, 98
Castor-oil plant, germination of seed of, 102–105
Cell, 41, 42, 48

Cell-sap, 43, 74
Cell-wall, 43, 114
Cells and tissues, 39–48
Cellulose, 41, 44, 106
Centrifugal force, 14, 15
Chemistry, in relation to plants, 23–31
— organic and inorganic, 30
Chlamydomonas, 114–123, 126
Chlorophyll, 29, 36, 53, 54, 59–65
— spectrum, 62
Chloroplasts, 53, 62–64
Cilia, 114
Coal, 26, 27
Coal Age, composition of atmosphere in the, 27
Colonial plants, 121–126
Compounds, 25
Conducting tissue, 40, 41, 45–47
Conifers, 94
Conservation of energy, 23
Cork, 42
Cotyledon, 5, 97–106
Culture solution, 35, 36

Darwin, Charles, 16, 18, 128, 130
— Francis, 16–18
Date palm, germination of seed of, 98, 99, 106
Decay, nature of, 87, 88
Diastase, 71
Dicotyledons, 100
Division of labour, principle of, 8, 40, 110–112, 122, 125
Donne, John, 77

Egg, of a plant, 3, 44, 45, 96, 107, 108
Element, 24
Elements essential to the green plant, 36
— in the soil, 34
Elm tree, number of leaves on an, 51, 63
Embryo, growth of plant from, 3–6, 108
— in seeds, 96–109

INDEX

Embryos of animals and plants, 4
Energy, 9, 10
— measurement of, 32
— sources of, 22, 30–38, 59–72
Enzymes, 71, 72, 106, 108
Epidermis, 51–55, 72
Euglena, 111–114
Evolution, early stages in, 110–122
— later stages in, 123–130
Eye-spot, 111–113

Fermentation, 71
Ferns, 94
Fertilization, 96, 125
Flagellum, 111, 112
Flowering plants, 94, 110
Flowers, 94–96
Food as a source of energy, 23, 30, 31, 33
— nature and source of plants', 33–37
Fruits, 100, 101

Geotropic curvatures, 16, 19
Geotropism, 16
Germination of seeds, 101–109
Gonium, 120–123
Gravity, stimulus of, 13–21
Grew, Nehemiah, 1

Heliotropism, 16
Hooke, Robert, 42
Humus, 74
Hydrogen, 25

Irritability, 37, 113

Kinetic energy, 23, 102
Knight, Thomas, 13–16

Leaf, the green, 49–72
Leaves, sensitiveness to light of, 50, 51
Leguminous plants, 89
Leonardo da Vinci, 49
Life, energy and, 23
— origin of, 42, 119
Light, chlorophyll and, 62
— nature of, 59–61
— stimulus of, 10–13, 113
Lignin, 47

Machines, plants as, 8, 22
Man, comparison of tree and, 2, 3
Maskell, Dr E. J., paragraphs contributed by, 55, 56, 77–80
Maxwell, Clerk, 22
Mechanical tissue, 41
Micrococci, 86
Mimosa pudica (Sensitive plant), 9
Molecule, 25
Monocotyledons, 100
Mosses, 94
Mummies, seeds from Egyptian, 101

Newton, Sir Isaac, 60
Nitrates, 84, 88, 89, 91
Nitric acid, 84, 88
Nitrites, 84, 91
Nitrobacter, 88, 89, 91
Nitrogen, 27, 37, 69, 83–84
Nitrogen-cycle, 37, 91
Nitrosomonas, 88, 89, 91
Nitrous acid, 84, 88
Nodules on roots, 89
Nucleus of atoms, 25
— of cells, 41, 107, 112

Organs, of plants and animals, 8, 39
Osmosis, 75, 76
Ovary, 95
Ovule, 95, 96, 97
Oxide, 26
Oxygen, 27, 33
— given off by green plants, 67–69

Palisade cells, 52, 53
Phloem, 7, 47, 66, 67
Pine tree, seed of, 105
Pistil, 95
Plant, chemical composition of, 33, 34
Plantain leaf, conducting strands in, 40
Plants compared with animals, 1–8
— common origin of animals and, 2, 43, 110, 113, 119, 129
— superiority to animals of, 22–31
Potential energy, 23, 33, 102
Priestley, Joseph, 68

INDEX

Primrose, flower of, 95, 96
Proteins, 31, 36, 38, 69, 107
Protoplasm, 3, 4, 37, 43–45, 74–76, 119, 120

Radiant energy, 29, 38, 59–72, 82; *see also* Energy
Ray, John, 100
Rennet, 71
Reproduction, methods of, 113, 116, 118, 122–127
Reserve food in seeds, 97, 106, 107
Respiration, 33, 58, 68
Rings of growth, 3
Root-cap, 73
Root-hairs, 73–75
Root-pressure, 76
Roots, 73–82
— effect of gravity on, 13–20
— their reaction to water, 19

Salts, 34–37
Sap, ascent of, 57, 76–82
Scott, E. T., 63
Seedlings, response to stimuli by, 11–20
Seeds, 70
— and seedlings, 70, 94—109
— germination of, 101–107
— longevity of, 101
Seneca, 129
Setaria, 11
Sexual reproduction, 117, 118
Sieve-tube, 46, 47, 65, 66
Soil, 74, 85
Spectrum, chlorophyll, 62
— solar, 60, 61
Spore, 94, 95
Starch, conversion of sugar into, 70, 71
— in leaves, 64, 65, 67, 69, 70

Starch in seeds, 102, 107
Stems, effect of gravity on, 13–15, 21
— structure of, 39–41
Stephenson, George, 83
Stimuli, internal, 20, 21
— response of plants to, 9–21
Stimulus, 9, 10, 20
Stoma (pl. Stomata), 52–58
Strengthening tissue, 45
Sugar as a source of energy, 65, 69
— convertible into starch, 67, 70
Sunflower, stomata in leaf of, 55

Tissue, 43–47, 127
Transpiration, 56–58, 76, 80–82
Trees, comparison of men and, 2, 3
— limit of height in, 3
— longevity of, 3
— structure of stem of, 7
Tropism, 16

Urea, 29, 87

Vacuoles, 45
Veins, in leaves, 50, 54, 66
Vessel, 46, 47
Volvox, 123–128

Water, composition of, 25, 26
— directive influence on plants of, 19
— giving off by plants of, 56–58; *see also* transpiration
— importance to plants of, 37, 101
Wheat, composition of grain of, 33
— mummy, 101
Wöhler, Friedrich, 29
Wood, 47
Work, definition of, 23

Yeast cells, 71

For EU product safety concerns, contact us at Calle de José Abascal, 56–1°, 28003 Madrid, Spain or eugpsr@cambridge.org.

www.ingramcontent.com/pod-product-compliance
Ingram Content Group UK Ltd.
Pitfield, Milton Keynes, MK11 3LW, UK
UKHW040157230326
469255UK00012B/148